Eat Like a Fish

Eat Like a *fish*

MY ADVENTURES AS A FISHERMAN
TURNED RESTORATIVE OCEAN FARMER

BREN SMITH

ALFRED A. KNOPF

NEW YORK

2019

THIS IS A BORZOI BOOK PUBLISHED BY ALFRED A. KNOPF

www.aaknopf.com

Knopf, Borzoi Books, and the colophon are registered trademarks of Penguin Random House.

LIBRARY OF CONGRESS CATALOGING-IN-PUBLICATION DATA
Names: Smith, Bren, author.
Title: Eat like a fish : my adventures as a fisherman turned restorative ocean farmer / Bren Smith.
Description: First edition. | New York : Alfred A. Knopf, 2019.
Identifiers: LCCN 2018050112| ISBN 9780451494542 (hardcover : alk. paper) | ISBN 9780451494559 (ebook)
Subjects: LCSH: Smith, Bren. | Marine algae culture—United States. | Shellfish culture—United States. | Marine algae gatherers—United States—Biography.
Classification: LCC SH390.5 .S65 2019 | DDC 338.3/714—dc23
LC record available at https://lccn.loc.gov/2018050112

Some of the names and identifying features of people and places in this book have been changed to protect privacy.

Jacket images: (boat) Ed Aldridge/ EyeEm / Getty Images; (seaweed) Mary Evans / The Image Works
Jacket design by Kelly Blair

Manufactured in the United States of America
First Edition

For Tamanna—

nurses and farmers belong together

Contents

~~~~~~~~~~~~~~~~~~~~~~~~~~~~~~

Eat Like a Fish

Introduction

I am a restorative ocean farmer. It's a trade both old
and new, a job rooted in thousands of years of history, dating back to Roman times. I used to be a
commercial fisherman, chasing your dinner on the high seas for
a living, but now I farm twenty acres of saltwater, growing a mix
of sea greens and shellfish.

I've paid my debt to the sea. I dropped out of high school to
fish and spent too many nights in jail. My body is beat to hell:
I crawl out of bed like a lobster most mornings. I've lost vision
in half my right eye from a chemical splash in Alaska. I'm an
epileptic who can't swim, and I'm allergic to shellfish.

But every shiver of pain has been worth it. It's a meaningful
life. I'm proud to spend my days helping feed my community,
and if all goes well, I will die on my boat one day. Maybe get a
small obit in the town paper, letting friends know that I was
taken by the ocean, that I died a proud farmer growing food

underwater. That I wasn't a tree hugger but spent my days listening to and learning from waves and weather. That I believed in building a world where we can all make a living on a living planet.

Fishermen must tell our own stories. Normally, you hear from us through the thrill-seeking writer, a Melville or Hemingway, trolling my culture for tall tales, or a Greenpeace exposé written from the high perch of environmentalism, or the foodie's fetishization of artisanal hook and line. When fishermen don't tell our own stories, the salt and stink of the ocean are lost: how the high seas destroy our bodies but lift our hearts, how anger and violence spawn solidarity and love. There's more edge to fishermen—more swearing, more fights, more drugs—and we are both victims and stewards of the sea.

So this is my story. It's been a long, blustery journey to get here, but as I look back over my shoulder, a tale of ecological redemption emerges from the fog. It begins with a high school dropout pillaging the high seas for McDonald's and ends with a quiet ocean farmer growing sea greens and shellfish in the "urban sea" of Long Island Sound. It's a story of a Newfoundland kid forged by violence, adrenaline, and the thrill of the hunt. It's about the humility of being in forty-foot seas, the pride of being in the belly of a boat with thirteen others working thirty-hour shifts. About a farm destroyed by two hurricanes and reborn through blue-collar innovation. It is a story of fear and love for our changing seas.

But, most important, it's a search for a meaningful and self-directed life, one that honors the tradition of seafaring culture but brings a new approach to feeding the country among the wandering rocks of the climate crisis and inequality. As fishermen and farmers before me, all I've asked for is a job that fills my chest with pride, a working life that my people can write and sing songs about.

I still miss being a commercial fisherman. But that's over now. Overfishing, climate change, acidification have forced me to change course. Now I have more in common with a kale farmer than I do with fishermen. My life is quiet, constant—working the same patch of ocean day after day for over a decade. I can't hang out in the same bars: What fish tales would I tell? Would I swagger into the Crow's Nest, turn up the lilt of my Newfoundland accent, order a stout, and tell a yarn about seaweed? "There I was in fucking flat calm. Reached down with my gaff, hooked a buoy, and up came my kelp, glistening brown wide blades. Fifteen feet. Longest I'd seen it in years. Yes, b'y, it was something to see."

I'd be laughed out of the bar.

ABOUT THIS BOOK

Writing this book was hard. My early years are fogged with drug-fueled violence and adrenaline, and I suspect drenched in over-the-shoulder romanticism. A life seen in reverse is an untidy affair. I struggled with structure. After much wrangling, I decided to weave together five concurrent strands.

First is my evolution from fisherman to ocean farmer. It was a difficult, emotional birth. I had to rewire my nervous system to new tempos of work, grow a blue thumb, hang out with odd breeds of people, even learn a new vernacular of food. It was a bumpy trip: my first brush with aquaculture left me disillusioned, and I've made many mistakes along the way to becoming a restorative famer, but in the end I landed on my feet.

The second strand is my rocky romance with sea greens. Like most Americans, I was skeptical about moving seaweed to the center of the dinner plate. Honestly, except for sushi, it sounded kind of gross. But I fell in love with a food lover, and she took me by the hand on a long journey of discovery. We met chefs specializing in making unappetizing food beautiful and delicious, learned about the lost culinary history of Western seaweed cuisine, and tested out kelp dishes on roofers and plumbers. In the book I've included a handful of recipes developed by Brooks Headley and David Santos, two of the most creative chefs in the United States, whose work points the way toward a delicious future.

The third strand is instructional: how to start your own underwater garden. It provides the basics for building a farm, seeding kelp and shellfish, and provides tips on farm maintenance and harvesting. It's not comprehensive, of course, but it might wet your whistle.

Fourth is my journey of learning. I had a long history of struggling in school, but yearned for a way to understand my life on the ocean within a larger context. So I trace my learning curve through the rise of industrial aquaculture and the origins of restorative ocean farming to the secret strategy to convince Americans to eat kale and the emergence of the regenerative economy. There were many surprises along the way. Who knew that the Japanese consider an Englishwoman the birthing mother of nori farming and hold a festival in her honor every year? Or that a shipwrecked Irishman accidentally invented mussel cultivation while trying to net some birds to eat? Or that McDonald's pioneered a seaweed-based burger in the 1990s?

Finally, there is my tale of passing the baton. This didn't always go well. I swam with the sharks of Wall Street, drowned in viral media, and failed at building a new processing company. But it was worth the trip, because out of the ashes came GreenWave, a training program for new farmers, partnerships with visionary companies like Patagonia in the era of climate change, and a new generation of ocean farmers to take over the helm and release me back to my beloved farm.

You'll also hear a lot about kelp in the book. On my farm, we've experimented with a few different kinds of seaweed, but sugar kelp has emerged as the most productive, delicious, and viable native species in my area. Most of the book will refer to kelp, but know that, every day, farmers, scientists, and chefs around the world are figuring out new ways to grow and use the thousands of vegetables in the ocean.

Though I include a lot about kelp, I've written very little about Asian cuisine. The history of seaweed use in East Asia is well documented, and I could never do it justice. What's surprising is the largely unknown parallel history—reaching back thousands of years—of shellfish and seaweed cultivation and cooking in the West. At times these two histories intersect, which I explore in the book, but I figured my job as a U.S.-based ocean farmer was to explore my own regional roots, and ways to cook and work with sea greens within the region I know.

Although I'm an ocean famer, I am not a fish farmer. The vast majority of aquaculture has entailed humans' trying to grow animals that swim. Recently, there have been major advances in the industry, but my take—which routinely drops me into hot water with the fish-farming crowd—is that the United States possibly chose the wrong path for ocean agriculture and continues to do so. Over the previous decades, there were countless opportunities to reflect on the unique qualities of the ocean from an agricultural perspective. If the nation had chosen to focus energies on growing restorative species such as seaweeds rather than jailing and feeding fish, we'd have a more sensible dinner plate today. We'd be feeding the planet while breathing life back into our seas, and protecting wild fish stocks while creating middle-class jobs.

There will be gaps in my story. For example, I am estranged from most of my family, a history that I will address with silence. They are good people; I could have been a better man. And I've worked many different jobs in my life. I've driven lumber trucks and sold my wood carvings on the streets of New York City. I've worked on community-organizing campaigns with coal miners, immigrants, and fishermen, and even did a stint for a politician. But these have all been mere tributaries. Though I have been pushed off the water many times in my life, I have always fought to return. This is a tale of a fisherman's forty-five-year search for meaningful work at sea. Maybe one day I'll spin another yarn of my dizzy days on land, but for now I'll stick mostly to the saltier side of my life.

A note about foul language. In his book *Distant Water: The Fate of the North Atlantic Fisherman,* Pulitzer Prize–winning author William W. Warner tries to prepare his readers for the cultural shock to come:

> The fishermen in some chapters swear more than others. No national slights are therefore intended. All fishermen have their choice epithets. . . . Their oaths and swear words are mere interstices—points of emphasis, like raising one's voice—devoid of literal meaning. The reader should so understand them, and take no offence.

So, before we dive in, let me apologize in advance. I write about the slimy, violent end of things, especially in describing my early years. And I swear a fair amount, always have, which has remained a sore point even at home.

Convention says I should repent and prefer the sober, inoffensive, and violence-free life—but I don't. The knife's edge has been good to me. Making the world a better and more beautiful place isn't about "softening" for the dinner crowd. It's about the granular hard work of fighting waves and rolling sleeves up tattooed forearms to work with nature. It's not about "domestication," it's about blue-collar innovation. So leave civility on the docks; hop aboard and revel with me in the profane. It tastes so good.

A few words about word choice. While some prefer the term "fishers," I use "fishermen" to refer to both men and women who fish commercially for a living. Many have come to this consensus. As Clare Leschin-Hoar, who has extensively covered the fishing industry, explains: "I've met many female fishermen . . . and 100 percent of the time, they have told me they like the term 'fishermen'; they don't want to be called fishers." She says women in the fishing industry see the term as a badge of honor. "They worked very hard to become commercial fishermen, and they want that respect." I follow their lead.

To refer to my farming model, I use "restorative ocean farming," "regenerative ocean farming," or "3D ocean farming." This signals my search for a new lexicon for sea-based agriculture. I hate the term "aquaculture," but honestly, I haven't yet settled on what to call my type of farming. Same goes for seaweed, which I refer to as "sea greens," "sea vegetables," and "ocean greens." In some of the more scientific sections, you might even see "macroalgae"—that just means seaweed, too. If you have better names, let me know; I'm keeping a list. Also, for all you land-lubbers: at sea, rope is called "line."

WHAT IS RESTORATIVE OCEAN FARMING?

Picture my farm as a vertical underwater garden: hurricane-proof anchors on the edges connected by horizontal ropes floating six feet below the surface. From these lines, kelp and other kinds of seaweed grow vertically downward, next to scallops in hanging nets that look like Japanese lanterns and mussels held in suspension in mesh socks. On the seafloor below sit oysters in cages, and then clams buried in the mud bottom.

My crops are restorative. Shellfish and seaweeds are powerful agents of renewal. A seaweed like kelp is called the "sequoia of the sea" because it absorbs five times more carbon than land-based plants and is heralded as the culinary equivalent of the electric car. Oysters and mussels filter up to fifty gallons of water a day, removing nitrogen, a nutrient that is the root cause of the ever-expanding dead zones in the ocean. And my farm functions as a storm-surge protector and an artificial reef, both helping to protect shoreline communities and attracting more than 150 species of aquatic life, which come to hide, eat, and thrive.

Shellfish and seaweed require zero inputs—no freshwater, no fertilizers, no feed. They simply grow by soaking up ocean nutrients, making it, hands down, the most sustainable form of food production on the planet.

My farm design is open-source and replicable: just an underwater rope scaffolding that's cheap and easy to build. All you need is $20,000, twenty acres, and a boat. And it churns out a lot of food: up to 150,000 shellfish and ten tons of seaweed per acre. Because it is low-cost to build, it can be replicated quickly. Best of all, you can make a living: one farm can net up to $90,000 to $120,000 per year.

Finally, the model is scalable. There are more than ten thousand plants in the ocean, and hundreds of varieties of shellfish. We eat only a few kinds, and we've barely begun to scratch the surface of what we can grow. Imagine being a chef and discovering that there are thousands of vegetable species you've never cooked with or tasted before. It's like discovering corn, arugula, tomatoes, and lettuce for the first time. Moreover, demand for our crops is not dependent solely on food; our seaweeds can be used as fertilizers, animal feeds, even zero-input biofuels.

As ocean farmers, we can simultaneously create jobs, feed the planet, and fight climate change. According to the World Bank, a network of ocean farms equivalent to 5 percent of U.S. territorial waters can have a deep impact with a small footprint, creating fifty million direct jobs, producing protein equivalent to 2.3 trillion hamburgers, and sequestering carbon equal to the output of twenty million cars. Another study found that a network of farms totaling the size of Washington State could supply enough protein for every person living today. And farming 9 percent of the world's oceans could generate enough biofuel to replace all current fossil-fuel energy.

FORK IN THE ROAD

In 1979, Jacques Cousteau, the father of ocean conservation, wrote: "We must plant the sea . . . using the ocean as farmers instead of hunters. That is what civilization is all about—farming replacing hunting."

Cousteau's dream—and mine—of hundreds of ocean farms dotting our coastlines is unsettling to some environmentalists, because it represents a new vision for our seas. I'm sympathetic to these fears, especially given the history of industrial aquaculture in the 1980s. But we face a trinity of crises: the leveling of agriculture yields, skyrocketing global population, and plummeting global fish stocks.

Unlike the climate change deniers sequestered in their congressional bunkers, these environmentalists don't dispute the facts, but the true significance, the implications, the urgency of the crises haven't yet sunk in. To keep pace with rising population, we need to produce as much food in the next fifty years as was grown in all of human history. But agriculture already uses 70 percent of the world's freshwater resources and produces more carbon emissions than all of the world's power plants combined. As a result, more and more of our food system is being pushed out to sea, and the UN estimates that 90 percent of world fish stocks are already maxed out under the stress of overfishing and climate change. In Asian waters alone, if the world continues at its current rate of fishing, there will be no fish left by 2050. That's only a few decades from now! Our wild fish cannot continue to bear this burden.

Necessity pushes us to farm the seas, but we can embark on our journey with anticipation and joy. With ocean agriculture still in its infancy, we have an unprecedented opportunity to build a food system from the bottom up. We can avoid the mistakes of industrial agriculture and aquaculture, farm for the benefit of all, not just the few, and weave economic and social justice into the DNA of the blue-green economy, all the while capturing carbon, creating millions of jobs, and feeding the planet.

Just in time, our seas are here to save us. As Jacques Cousteau said: "The sea, the great unifier, is man's only hope. Now, as never before, the old phrase has a literal meaning: we are all in the same boat." Indeed.

This is our chance to reimagine our dinner plate by inventing a new "climate cuisine," not around our industrial palate of salmon and tuna, but around the thousands of undiscovered ocean vegetables and shellfish found right outside our back door. Picture hundreds of small-scale ocean farms dotting our shorelines, surrounded by conservation zones supporting wild fisheries and breathing life back into our oceans. A Napa Valley of ocean merroirs, producing ocean vegetables with distinct flavors in every region. Ocean farms embedded into wind farms, harvesting not only wind but also food, fuel, and fertilizers.

In 1962, President Kennedy reflected on our bond with the sea:

> All of us have, in our veins, the exact same percentage of salt in our blood that exists in the ocean . . . Salt in our blood, in our sweat, in our tears. We are tied to the ocean. And when we go back to the sea . . . we are going back to whence we came.

The time has come to return from whence we came. What a beautiful tale this could be about the return of a prodigal nation. We were founded as a maritime nation; more of U.S. territory is located underwater than above. Every other breath we breathe comes from ocean ecosystems. If the pioneering spirit of the nineteenth century was captured by the instruction to "go west, young man," then this book is a twenty-first century call for our generation to "head out to sea."

part *I*

CHAPTER 1

Thank God We're Surrounded by Water

I was born in Maddox Cove, a Newfoundland out-port bolted to the easternmost rocks of North America. A couple dozen houses, a fish plant, and a fisherman's co-op, all brightened with coats of leftover boat paint—oranges, reds, yellows, greens—guiding us home through the fog.

Our house, a traditional two-story saltbox, was perched on a hundred-foot cliff. It was a sometimes brutal, sometimes beautiful life. Dad believed in hard, heatless living: frozen toilet water in the morning, and oversized down jackets from Canadian Tire. Mom, born in Brooklyn, New York, was delicate and rail-thin, and spent decades shivering against the wind.

Both Americans, they had crossed into Canada in protest of the Vietnam War. They were in their early twenties choosing between Newfoundland and Montreal, both places with their own promise of romance and adventure. Newfoundland won out, a tack that shaped the course of my life. I might have grown up nibbling croissants and sipping espresso, but instead I was born on the Rock.

Newfoundland became my father's domain. He was manic and larger than life, and secondhand stories about him litter the outports. The time he cut a house in two with a chainsaw, pulled each half by truck over a mile, and nailed it back together. Or when his boat engine died and he was marooned among the towering icebergs and paddled home. When he ate raw baby seal. When he stopped the drunken slaughter of a pig by a gang of kids who were chasing the poor swine with sledgehammers. And the time he built a barn with no nails and no saw.

A linguist who studied with Chomsky, he spent months at a time living and working with the Inuit of Labrador. Year after year, he would take the mail boat up the coast, sleeping on the deck, living in the grinding poverty of the tundra. He was a father who lived with the intensity of the true North. He would return home in a swirl of stories: hunting for seal on the ice pack; dogs eating a four-year-old child; his good friend Ken, the bear hunter, shooting at American military jets when they scared away caribou.

Back in Maddox Cove, my mother was often on her own with two small children. She had to learn to keep house at the edge of the earth. There were no fresh vegetables, so she planted a garden, but was only able to coax potatoes, cabbage, and turnips from the unforgiving rock. She hauled firewood to heat the house, churned butter by hand, and baked mountains of rolls. All the while, she had to be vigilant to keep my sister and me from tumbling over the cliffs. She kept Newfoundland custom, serving fish cakes on Fridays, and looked the other way when neighbors plied me with soda, candy, chocolate bars, and French fries. My mom was cut from different cloth than my dad, but they shared a fearless, open spirit.

Our house was crammed with collected treasures: half-broken antiques that my dad was going to fix one day, huge pieces of driftwood that my mom hauled back from the shore. Dad drove Mom crazy by taking apart engines in the living room, leaving parts scattered everywhere. And there were towers of books. Books about statistics, Cuba, poetry, Sartre. Books of riddles. A book about the Mafia, another on the Japanese game of Go. Engine-repair manuals, mushroom-hunting guides. Eventually it got so bad that they built a two-story-tall bookcase, firmly bolted to the wall, that was designed to be climbed to retrieve a volume. That was our jungle gym.

In the attic, there were dozens of boxes of cassette tapes. I'd sneak up with my tape player, choose one at random, and listen for hours. I'd hear disembodied voices from worlds apart. Inuit and Gaelic songs. Newfoundland outport stories and reels. Bits of German, French, Italian. A lifetime of hoarding for a linguist.

Over time the family grew, but not in a straight line. My dad set up a homespun work-release program, recruiting Inuit parolees from the local jail as graduate assistants for the Inuit dictionary he was working on. They lived with us, and their stories became part of the fabric of our family. Nora stuffed my cheeks full of warm cakes two at a time, and made me laugh by dislocating her jaw and pulling her lips over her cheeks and nose. Grace was a quiet, steady hand, helping my mom keep me in line; she went on to become a well-known scholar in her own right.

Nora and Grace had the haunting gift of laughing at pain. Not at others' misfortune, just at their own. The tail ends of their stories of beatings by boyfriends or the bigotry of neighbors were punctuated with full-throated belly laughs. I learned to giggle along. And the trait soaked in. Decades later, upper-class New England WASPs would point out my dark habit, worried I had a fetish for pain. I don't. The Irish sum it up: Gather around you those that have just the right amount of suffering.

When my parents worked in the yard, my sister was old enough to play on her own, but I had to be locked in the "whale pen"—a roll of fishing net staked out in a circle—where I spent hours throwing myself against the sides with all my baby-fat fury, unable to reach the cliffs just beyond my cage. An extreme stutter kept me mute and fiery with rage. The doctors advised teaching me sign language, but Mom pushed hard to unlock my tongue, spending hours playing my favorite card game, Go Fish, while practicing impossible sounds.

I retreated to the sea, combing the pebble beaches for whale bones and sea glass. The rolling breakers, sharp relentless winds, icebergs floating outside the harbor—this was my comfort zone. One day, in Middle Cove, my mom found me building a wall of beach stones and chattering fluently to the sea. She listened with surprise. The ocean helped loosen my tongue, and I've never shut up since. The only time in my life my stutter returned was while dropping acid in my teens.

As my mom had learned, growing food on the Rock was all but impossible. Vegetables and fruits were shipped to our shelves in the belly of the Joseph and Clara Smallwood ferry, arriving mushy and discolored. Meat was long frozen, passed from province to province, till it hit the docks with freezer burn. The national dish was the Jiggs dinner: a pot of cabbage, turnips, and potatoes simmered for hours, with, on rare days, a hunk of salt beef. It was poor folks' food, but well loved.

Years later, on a return trip to Newfoundland, I took an old couple out to dinner at a fancy restaurant on Water Street in St. John's. They hadn't been out to eat in nearly a decade and ate Jiggs many nights. They pored over the menu, eyes wide at the seared salmon and prime rib. When the waitress came to the table, the old woman ordered the Jiggs dinner. I had noticed the back of the menu offered up "Favorite Newfoundland Dishes" for the tourists to play pauper for a day.

With a mix of frustration and disbelief, I asked, "What the hell are you doing, ordering Jiggs dinner?" She looked at me with a sly smile. "Well, I wanted to try theirs!" Only a Newfoundland cook can take the mainland's rotting produce and turn it into cherished cuisine.

Seafood was a different story. Fish was measured in hours, not days. Shopping for dinner meant a five-minute walk down to the docks for cod, flounder, or bycatch coming off the day boats. Kids sold cod tongues door-to-door for pocket change. When the squid came ashore by the thousands in the spring, we'd run down to the beach to turn them into squirt guns. I'd grab one slippery six-gun in each hand and give a hard squeeze, squirting blue-black streams of ink into my sister's hair. We'd eat them, too, piling driftwood and rocks to roast squid strips till they charred and curled. A flask of Ballantine's would soon slide from someone's pocket, and if the weather held, an old-timer would show us how to hammer out a few reels on the fiddle.

Our neighbor Mrs. Doyle would trade pickled moose for a jar of Mom's partridgeberry or bakeapple jam. Hardtack—two-by-four-inch blocks of flour, lard, and salt—was the fisherman's midday snack. It could break a tooth, but had the shelf life of a Twinkie and softened to sweet mush in a cup of tea. On weekends, Dad and I would head to the cliffs of Middle Cove to hunt for mushrooms. We'd race through the moose trails—Dad did most things at a half-to-full running speed. He'd point to a spruce tree, and I'd drop to my knees, stick my head under the boughs, and pop back out with a bunch of chanterelles. Back home, cooking them in a pan of butter, with the smell of mushroom meat hanging heavy in the air, he'd repeat stories about European tourists who died violent, convulsive deaths because they confused Newfoundland species with their own. As I shoveled in mouthfuls, I'd imagine emergency rooms packed with mouth-foaming tourists strapped to hospital beds, unable to say goodbye to their families.

We were 1980s foodies. When McDonald's hit the stony beaches of Newfoundland, hundreds of us waited in line. We swarmed from the nooks of our rock: Bay Bulls, Goulds, and Flatrock. Chatting about the future of our island, what we'd order for our first meal. My first time, I choked down two Big Macs, large fries, and a vanilla shake. It was the best thing I'd ever tasted. To this day, when I need a dose of solitary pleasure, I pull into McDonald's to order a few Filet-O-Fish sandwiches. This need will never die.

My great-aunt, a lifelong secretary at Standard Oil in New York City, used to mail us monthly packages of toilet paper, paper clips, pencils, and note cards. I still have the fifty-cent pin the multibillion-dollar company gave her on her forty-fifth work anniversary, which she had made into a ring and wore with pride. Every couple years, we'd make the long journey to the States to visit. Off-island, I ate in a panic. After the nine-hour drive and seven-hour midnight ferry to reach the mainland, my parents would pull into an all-you-can-eat chicken place. I ran from the car as fast as I could, leaving my family behind. By the time they came through the doors, I'd already be tearing through a plate piled with wings.

No surprise that hockey ran bone-deep. Like many kids north of the border, we woke at 4:00 a.m. to practice, and every weekend was taken up by driving game to game. The first non-kid book I read was the autobiography of Wayne Gretzky (my second was a bio of Muhammad Ali given to me by my dad, which I had trouble wrapping my Newf head around). I was pretty good, playing on a team of kids two years older than me, which meant I got to wear an orange jacket with black stripes to school. I was small but disciplined, running drills and practice pucks on every patch of ice I could find.

My favorite hockey game of all time was between the New York Rangers and the Montreal Canadiens. Packed rink, third period, both the crowd and players itching for blood. Down by one goal, and the announcer comes on: "If we score a goal in the next two minutes to tie it up, everyone gets a free hamburger." We leapt to our feet, fans on both sides, and began to stomp and chant, "Hamburger! Hamburger!" The stands and glass shook. "Hamburger! Hamburger! Hamburger!" The center forward took a shot from the outside, hit the top right corner of the net. Scored. We went crazy.

As an outport kid, I was bused an hour into the city for school. I was built for the yard, not the classroom. We played marbles and brawled at recess in a playground with no grass, just dirt and gravel mixed with shards of glass. Rain or shine, I'd lie belly-flat in the mud in my tight white T-shirt and brown cords to line up shots of balls ranked as "goat's eyes," "normals," and "ball bearings." Soon I was banned from classes, because I'd sneak back in covered in filth. My mother was called and told I'd be kicked out unless I cleaned up. She knew the pent-up fury I carried into the classroom, always struggling to follow what others breezed through with ease. Her solution was to bear the burden: each day, at lunchtime, she'd drive an hour to school, bringing hot towels, a change of clothes, and a Swanson's Chicken Pot Pie. I'd climb into the back seat, strip, and let her wash me down as I stuffed in bites of warm crust, gravy, and meat. A mother's love runs deep.

Though classroom grammar and numbers left me isolated and confused, the lessons of the outport did not. We learned at the edge of truth and untruth, where the magic of the cliffs met the raging of the sea. We were taught that injustice must be met with swift and righteous punishment from the story of how the local priest, a massive barrel-chested giant, beat a fisherman to a cowardly whimper for drunkenly beating his wife. We learned the value of blind compassion from the story of how Mr. Doyle, who couldn't swim, jumped off the pier to save a dog, and then had to be fished out before he drowned himself.

We were a people that salted our words. "Who knit ya?" meant "Who are your parents?" "Fire up a scoff" meant "Make some food." "Gutfounded" meant "hungry." "Long may your big jib draw" meant "Good luck." "Put the ol' slut on" meant "Put the kettle on the stove." Television channels used subtitles when our people were interviewed on the nightly news. Buying was a protracted affair. Cultural norms required conversation before commerce could commence. You'd go into the corner store, ask how things were going. Let the owner pass a bit of gossip or a quick health update of a loved one. You'd chat it up, wait for the owner to pop the question: What do ya need?

History lessons were homespun tales. We were once a self-governed land with our own foreign policy and currency, until a rigged vote forced a marriage with Canada, consecrated on April Fool's Day, 1949. Families floated their houses across the water when the government shut off supplies in the name of progress—more than 250 outports were emptied between 1954 and 1975. Foreign companies stole our fish while we read books like *No Fish, Our Lives*. The best-selling album throughout the 1980s had a picture of an oil platform on its cover and was called *Original Newfoundland Gold*. And surely we were the only country whose nationalist anthem contained verses about "blinding storms," "wild waves," and "tempest roars." Our favorite bar sing-along was "Thank God We're Surrounded by Water."

Firemen, cowboys, astronauts: those were off-island dreams.

Where I grew up, it was fishermen who filled my silver screen. They worked in the event horizon, out of reach, and barely governed by the rules of men, and when they hit the docks, laughs and whoops announce their landless happiness.

The allure of the fisherman's life is the anticipation of each haul—that leaning over the gunnel, not knowing what might be pulled from the sea. Even after decades of fishing, that moment still thrills me: the sound of the hauler straining under weight, the first glimpse of fish as the net breaches, then the delight or disappointment at what's inside. Maybe it's the sheer volume of fish, how mind-blowing it is that so many things that swim can be pulled from the sea. Or the sheer diversity hidden in every ocean: sea cucumbers, octopuses, snails, stingrays, jellyfish, corals. And then there's the haul of men: lost anchors, traps and cages, gloves, and bottles. Every fisherman has a story that captures that moment when the nets break the water. Once, a fisherman told me a tale of the time he pulled a missile up in his nets and the Coast Guard cleared the harbor as he came into port.

The first fish I caught was a cod. I was probably four or so. My father took me out to Flatrock, a huge slab of sandstone that tilts into the sea. Weather started fine, but switched suddenly to sideways rain. Cast after cast, no bites, just snags. To keep warm, I sipped tomato soup from an army-green thermos my father had packed. It was getting dark, so he told me, equally disappointed, the next was my last cast. Out went my line. I reeled in slow to delay our leave. Then a hard yank, and I landed my first fish. I have a picture of my father and me in the kitchen, holding the huge cod. It was half my body in length, and there was a new hunger in my eyes.

CHAPTER 2

Saltwater Cowboys

J ust as I was almost old enough to skipper a skiff, my parents shanghaied me off the docks and forced me into a rigged referendum. Dad called us into the living room and told us that he and Mom had been talking and thought it was time to head to the States so we could get into better schools and my mom could find a job. He called for a show of hands—whether to stay or go. For me this was a choice between staying in a glimmering world of sea glass, hockey, and cod tongues, or heading for what I thought was a landlocked tooth-and-claw of a country. I lost, three to one. We piled our things into our orange station wagon, and I said goodbye to the sea.

We moved to suburban Massachusetts. I was an innocent-eyed Newf kid—good at hockey and hooking cod. On my first day of school, I dressed up to make an impression—rust-colored corduroys, a striped shirt, cowboy boots I'd gotten for my birthday. The suburban kids were ruthless. When I talked, they laughed, unable to understand my Newfoundland accent. It didn't help that I looked like a 1970s Canadian cowboy.

Plus, I couldn't keep up with the schoolwork, and at home, things weren't much better. Since we arrived in the States, my parents' marriage had been in a bad place, and the family was falling apart. I retreated, sliding down the white working-class ladder. At school, I started getting into fights—first with other kids, then with the principal, finally with the cops at a football game. The school and I agreed: it was time for me to go.

My girlfriend Stacy and I, locked in a downward embrace, dropped out on the same day and moved into Section 8 housing with her mom. Her mom got us both jobs at the local hospital where she worked, Stacy in the X-ray department, me as night-shift janitor in the emergency room. I was fourteen and mopping up piles of vomit and blood for a living. Pushing a supply cart, changing bedsheets, scrubbing toilets, emptying trash. I wouldn't see my own family again for months.

The hospital was up the road from a prison, so pushing a broom was high-octane work. Inmates, handcuffed to their beds, squirmed in pools of blood. One time, a farmer stumbled in to the ER, his pants soaked in blood. He had jumped off his truck while holding a pitchfork, and the fork had stuck into the dirt so that its wooden handle impaled his scrotum up through his innards. He pulled it out and drove himself to the emergency room. Another time, an old man came in after a stroke, has face half paralyzed. I scrubbed the bed next to him, a curtain between us, while listening to him trying to eat his breakfast. It wasn't working. He'd fork a bite, try to swallow, vomit. "Fucking coward," he'd growl and try again, spooning his vomit. Tough bugger of a man—wouldn't surprise me if he was a farmer too. I moved on. He needed to fight this battle alone.

The ER nurses wouldn't let me be. The first day I pushed my cart into their domain, they swarmed, demanding to know my story. They were all ages and sizes, but all shared a mix of warmth, strength, and irreverence. They made fun of my teenage silence and anger, forcing me to unfurl. I'd join them at the nurses' station, listening to gossip about new patients, shitty boyfriends and husbands, grim tales of bad sex, frustration with sloppy and arrogant doctors. They were great storytellers, able to slide from ridicule to compassion mid-sentence. During that time in my life, they were my ballast. No wonder I later married a nurse.

Off the job, I was getting arrested a lot: stealing, dealing, fighting. Even went on the lam for a week, running from the cops after getting fingered for stealing. Too young to drive, I hid under a highway bridge for two nights, shivering in the late-fall freeze. The police were on the prowl, stopping my friends on the streets to find out where I was holed up. On night three, I slept outside a gas station. The temperature dropped below freezing, so I broke into a car and built a fire on the floor of the back seat, kindled by newspaper, a glove, paper cups. I could have blown myself up, but it got me through the night.

After a few days, I made my way back to Stacy's house, wading through the marshes, scurrying across backyards and roads after dusk. I hid in the attic. The police came to the house again and again to question her family. On the seventh day, her brother drove me to the police station so I could turn myself in. I was already on probation for assault on a police officer two months earlier. I look back on myself—that stupid, violent boy sitting alone in jail. I logged concussions from fights lost, and I still have scars on my hands from teeth crushed against my fists. There was an uneasy, unrelenting momentum to my life. That night, sitting in my cell, I figured before long I'd be dead or in jail for good.

These were the three-strikes-and-you're-out Reagan years. Black kids were being thrown into jail by the thousands, but I was playing the role of a white trash suburban kid, so I got pushed back onto the streets. The judge said I deserved a year in juvie, but my parents, who hadn't seen me in months and couldn't find a trace of their little Newf in my eyes, pleaded with him. He gave in, and I was sentenced to a regimen of NA meetings twice a week, AA three times a week, once-a-week sessions with my probation officer, and court-ordered therapy.

I was still too young to have a driver's license, so I hitchhiked to all the meetings; often my rides got me high. I'd arrive at NA group meetings in a fog. I was the youngest and least experienced of these hungry ghosts, so after the meetings they mentored me in artisanal techniques like dropping acid directly into my bloodstream through tiny cuts.

After NA, I'd hitch to work, pull my shift, crash, then hitch to an early AA meeting, where I learned from the living dead to drink black coffee and chew tobacco at the same time. Long-winded, stupid, boring stories from the hoarse, wilted, unemployed, and destroyed. I still can't shake the smell: cigarette smoke curling with hopelessness. I sat in the back, refused to talk; no one cared.

I lived to work. On the job, numbness faded and the electricity in my blood watered down. School, AA meetings, police: all of it injected fury, revolt, a cruel streak into my veins. But in work—the routine, the physicality, the solidarity—I found a cure. This was my first job, and it was the first time I felt the emotional effect work has on me: the rest of the world fades, the chaos in my brain settles.

During my second stint in probation, I began spending more time with my mom, who was pushing hard for a new career path. She tried to get me a job as a fireman in Brooklyn, through some friend she knew. No one would take me. My court-ordered psychologist gave me an IQ test, which I filled out on the night shift at the hospital, marking the multiple choice questions on my janitor's cart. On the elevator, a doctor looked over my shoulder and asked me what I was doing. "Taking an IQ test," I mumbled. He stared at me for a bit, clearly unable to get his head around why some janitor was testing between scrubbing toilets. I was just as confused as he was. The puzzle pieces of my life weren't fitting together. I did well on the test, and my mom begged me to go back to school. But I had different plans. I knew I didn't want to push a mop my whole life, but school had never worked for me.

I took a job with Stacy's uncle, who ran a lobster boat fishing out of Lynn, Massachusetts; back then, this was one of the roughest towns on the East Coast—"Lynn, Lynn, City of Sin." Every morning, I got up at three, stopped in for a donut with the old salts, and was on the boat by four. I began my first week of work in the sweltering heat of August. She was a thirty-two-footer with a narrow beam, small for the weather we hauled in, run by Captain Mike, who liked a bit of cocaine with his coffee and had a glint of hate in his eyes. It was just the two of us on the boat. That first week, the seas were dead calm. The stench of the bait, a maggoted soup of rotted mackerel, pushed up into my nose and coated my throat. For four days in a row, on the drive home, I had to pull over and puke from land-sickness. That same week, I fell asleep at a stoplight and crushed the trunk of a BMW with my truck—thank God she didn't call the cops, since I only had my learner's permit.

I had barely gotten my feet wet, but, by the end of that first week, I fucking loved that job. Smelling like bait, salt, sweat, I no longer belonged onshore. Captain Mike worked me hard, hauling hundreds of traps a day. We listened to thrash metal. I saw every sunrise, every sunset. I was hungry all the time, scarfing whole bricks of cheese and loaves of bread.

My work was written in my hands: constant infections from rotten bait juice that seeped into gashes made by slips of my knife. Off the boat, they dried and cracked, bleeding whenever I spread my fingers. No girlfriend could bear my clawlike caress—it was like having sex with a lobster. My landlubber friends began drifting away as my edge sharpened. I hung out exclusively with fishermen after work. Mike and I were teetering close to friendship, hanging out, getting high, shoveling down meals together. It was good for me. No matter how raw and misguided my mentors, I desperately needed adult supervision.

One Saturday, me and the boss took the day off to go check out the latest lobster gear at a commercial fishing show set up in a warehouse in Gloucester. We showed up late, and when we walked into the main convention hall, all the fishermen were crowded around one booth. We pushed our way to the front of the crowd. Two women in bikinis were posing next to a stack of lobster traps, while a big-bellied guy was explaining the latest advances in trap design. We were a predictable bunch. My captain and I hung out and enjoyed the show, nodding along like we were listening.

These were the years of the lobster wars. Unless you fish—fish for money, I mean—it's hard to explain the logic of lawlessness at sea in the 1980s. I'm not talking about the shadow pirate fleets off the coast of Africa or the Thai slave trawlers. I'm talking about the postcard day boats floating in local harbors that, once they round the breakwaters and hit open sea, enter into low-level guerrilla warfare.

Mike had only been out chasing lobster for a couple years before I showed up, so the Lynn fleet was still trying to push him out. Lobstering was big money then, and the seasoned captains didn't appreciate newcomers poaching their profits. We were running eight hundred traps in trawls of forty traps each, all strung together with a buoy on each end. We'd set our lines, let them soak for two- or three-day rotations, and return to find them gone. Each time we lost a trawl, it was a couple thousand dollars overboard. Cut enough trawls and it drives a captain out of business. So you go to war: trawl for trawl. We'd always have a suspicion of who was cutting but were never sure. There's no way to catch someone red-handed, so make a guess and pull out the knife. The sea hides the crime—no evidence, no witnesses. Land law is neutered once you leave the harbor. Call it a good year if you don't have to pull a gun or a knife. Welcome to the untamed, ungovernable sea.

That winter was the first season when I was almost killed at sea. There would be many more, but the first time always sticks in the skull. During a week of deep freeze, we'd been banging ice off the rails all morning. It was blowing, and the swells grew into walls of water. Our thirty-two-foot lobster boat was open stern, meaning the stern of the boat was open so traps could slide off the back during the set. I was midway through a haul, cages sitting on the deck strung together into a trawl. Mike was setting into the wind and said that after we dropped these last traps it was time to head home.

It's always the last trawl when shit happens. You're tired, captains are getting sloppy, the edge begins to dull. There was ice everywhere, including on my eyelids and brows. My head was down when Mike yelled, "Fuck, look out!" and I looked up as the wave barreled down. It swept the deck and dragged the traps, the line, and me with it back toward the open stern. I lunged to grab the gunnel and pulled the knife strapped to my chest to cut the line. Barely made it—and loved it. The sea's lack of mercy was thrilling.

I caught this fishermen's disease early on. Best description I've run across is from a Coast Guard captain who diagnosed it in a letter to the editor of *The New York Times*:

So many commercial fishermen [seem] to be trying very
hard to die. I'm beginning to think there is a disease that
is caught early in a working fisherman's life; it's as if there
is something in the scales of fish that wants to pay them
back, something that gets under their skin. Once in their
blood, it affects the brain and makes them more likely
to die than any other group of professional mariners. It
makes them believe that they are different; that fishing is
more dangerous than every other job out there, and nothing can be done about it. . . . That belief causes them to
not even try to escape the danger, and they end up taking
risks that other professional mariners successfully avoid
every day.

To this day, I still haven't shaken the bug: I've never worn a
life vest, never learned to swim. Many of us born up North don't
know how—swimming just prolongs the inevitable. Out all winter alone, sliding on icy decks, hanging off the gunnels . . . I still
collect stories of close calls to send shivers down the spine of
my wife. The best advice I ever got on the water was from an
old salt who gurgled and wheezed when he talked because of a
poorly patched bullet wound. Growling more than speaking, he
warned me, "Once that fear leaves you, boy, get the fuck off the
water." Best advice I've ever ignored. It's just in my blood—I dare
the sea to take me. I hope one day it does.

While I was still working in Lynn, the IRS showed up at my
mom's door. Turns out I hadn't been paying taxes. I thought
I was getting paid under the table, but Mike had reported my
earnings to the government. I was making good money for a
sixteen-year-old kid. I had to sell my beloved motorcycle, my
truck, and even a bunch of my tools to pay the fines.

As a small side gig, I started reselling "bugs." On the way

home, I'd stop by the office where my mom worked as a secretary. Still in my rubber boots and Boston Bruins hat, smelling of sweat and mackerel bait, I'd sell live lobsters, cubicle by cubicle. My mom's co-workers would be ready for me, coolers tucked under their desks. I'd lay a couple wet "bugs" on the carpet, and they'd squeal with delight, picking each one up, testing to make sure there wasn't too much flex in the shell, a sign of scant meat. I gave them a break in the monotony of their workaday world—a break in the alienating commercial food chain. They could have gotten the same lobster at the Big Y Shopping Center, but this felt like a return to some natural order.

This was the late 1980s, the decade of Gorton's Fish Sticks and TV dinners, and I was just trying to hustle some money on the side to fund late nights of teenage bourbon and blow. But the embers of Maddox Cove still glimmered in my chest. The little salted runt of a Newf from a town where kids sold cod tongues door-to-door was now sixteen years old, running an underground sea-to-cubicle pipeline into the gutless office parks. This was my first taste of the pleasure of feeding everyday people. Even as a teen, I'd get high thinking of the secretaries and mid-level managers heading home with coolers of my lobster riding shotgun in their Toyotas.

CHAPTER 3

Alaska Bound

After a couple years of lobstering, I got the itch for more. I was seventeen, and the hard-living, hard-drinking years on the docks drove me toward rougher seas. Fishing in Alaska was getting a shot at the brass ring for an up-and-coming fisherman. With no particular plan, I hopped a plane to Seattle.

From Seattle, I hitchhiked the twenty-six hundred miles to Anchorage, along the way perfecting the culinary art of the cost-cutting condiment sandwich: two pieces of Wonder Bread smeared with ten packets of ketchup, mustard, relish, and sugar lifted from gas stations and fast food joints along the way. Hitching up to Alaska is a breeze. Truckers, hunters, acid freaks run the trip nonstop. Just wave them down and it's a free ride straight to Anchorage. Drivers expected me to smoke, drink, gobble mushrooms mile after mile—and take over the wheel when following yellow lines became impossible for them.

My in-kind donation was knowing how to drive under the influence—a skill mastered before I owned a driver's license.

Along the way, I met the vagabonds of the Alcan. The long-haul trucker who bragged about sleeping with the "ugliest women in the world"; a gaggle of Alberta hunters who were so hard-drinking that I demanded they pull over to let me out on a barren stretch of highway. Then there were the father and son who had just hit gold in a mining claim they had been digging for years. Before I had even settled into my seat, they told me to hold out my hand and dumped dirty nuggets into my palm. It all had a clichéd *On the Road* feel: hot springs with sagging hippies; humbling mountain views; one-night stands with liberal arts college girls hungry to return to campus with sweaty tales of fucking the underclass.

From Anchorage, I hitched a ride to the Spit in Homer, pitched a tent, and lived on the beach for a few weeks. There weren't many of us, maybe a dozen, and I quickly figured out the Spit wasn't my scene: too many dreadlocks playing tin whistles around the fire. None of them were there to fish, and certainly not to spend their lives at sea. They were there for the mountains, the dope, the summer music. I was there to work, learn my trade. These weren't my people.

I got a short stint working in a cannery in Naknek, the center of Bristol Bay's salmon fishery—stuck on land. It was a boring job that couldn't end soon enough: eighteen-hour shifts, slicing guts from salmon along an assembly line of college kids and Inuits, zero thrill. The only fun was gut fights across the lines. One time, I made a snowball of intestines and threw it at a girl I wanted to get with. She yawned just as I chucked, and it landed in her open mouth. She puked and hated me. When the job was over, I returned to the Spit to wait.

Maybe the idleness was the trigger, but this is when I began experiencing hauntings. Every time I closed my eyes to sleep, a face stared back, nose to nose, from behind my eyes. It was a man, maybe in his forties, bearded and stern. He was a constant hovering ghost at night. I lived with him for months, and then, one night, he didn't show up. I've never seen him again, but it marked the first of many odd tricks my mind would play over the coming years.

One morning, a big hand rattled my tent. I poked my head out and found a grizzled beef jerky of a man, who shouted, even though it was a bone-quiet morning: "No one on this fucking beach wants to work. How about you? Wanna coil some line?" I jumped out of my tent.

He drove me to a lonely field where there were thousands of feet of rope and stacks of black totes. "Coil that line into the totes. I'll be back later." He drove off. I knew how to coil, and took pride in my carefully but quickly laid spirals, so I pulled up a log and did what I was put on earth to do: work. I'm not one of those "Zen and the art of work" fools, drifting into a meditative stupor of deep thoughts, but the world gets more vivid, less sentimental when I'm on the grind. I was born with really only one talent: to work continuously for hour after hour. I coiled and coiled, for four, six, eight hours. The boss stopped by, looked at my work, and asked, "You wanna stop?"

"No," I said, "I'll keep going." He drove away. I worked another six hours, into the evening daylight of the Alaska summer. He came back. Walked over. "You want a real job?" This had been a test to see if I could put in hours without a whimper, and apparently, I'd passed.

Next day, he flew me into Dutch Harbor, the lodestar of the Bering Sea. This was twenty years before *Deadliest Catch* turned Dutch into a TV fantasyland and a walkabout for English majors. When I showed up, the Elbow Room was rated the most dangerous bar in America by *Playboy* magazine. I had been in training for the underbelly of commercial fishing for years. I could drop acid, follow with a spike of cocaine, drink all night, and still hold my own in a fight. I'd had knives pulled on me, was on my fourth concussion, and got invites to Hell's Angels weddings and funerals. So Dutch was my kind of town.

A week after I landed in Dutch, the DEA landed in force, busting "fishing" captains for smuggling heroin in from Russia. This is a part of the story that's often left out. I've heard rumors that the industry has mellowed over the last few decades, but this was the 1980s when there was heroin hidden in the herring. Years after I left Dutch, I found myself at dinner with a tableful of ocean economists, forced to listen to them drone on about catch shares and shrimp commodity prices. I finally elbowed my way into the debate, telling tales of how drug running helped pay the bills for many a fishing captain. They were horrified; I suggested they conduct better fieldwork.

Even though I was still a kid, I had already been fishing for years, so jobs in Dutch came easy. I worked trawlers, longliners, and crab boats. The work was hard, but I'd been forged for the twenty-to-thirty-hour shifts for weeks on end. We were a mash-up of Inuit, white trash, Mexicans. It was the first time I met someone with a tear tattooed on his cheek, a badge for killing his first man. I learned how to sleep standing up, and how to spike chewing tobacco with jailhouse hooch fermented from cans of fruit cocktail. I can still hear the grunting sound cod make when you stun them with wooden bats.

My first boat was a 170-foot longliner. It took me a few days to get my sea legs back. I worked in the belly of the boat, in the gear room, twenty-hour shifts with three hours to sleep. We'd set forty thousand hooks at a time. I ran the squid-bait feeder. The first week, I was adjusting slowly to the long hours and bouts of nausea. Few admit it, but it's common for fishermen to spend the first few days seasick as we adjust to the thunder and roll of working in the belly of the ship, the air thick with the stink of engine oil and rotten bait. A low point: early morning, and we had our heads down cleaning gear for hours, death metal playing on a loop. I headed to the bathroom and sat down on the toilet. Sound was coming through in waves, and the walls seemed to be dripping. I fell asleep for a minute or two, began to stand, and realized that I had shit on the seat. I sat back down in the filth, fell back asleep, woke, pulled up my pants, and headed back to work. The nostalgic pull of fishing is so strong that I think back even on this as a fond memory.

We were fishing the summer waters of the Bering Sea, closer to Russian shores than to American. We'd been out for two months, trawling ocean ledges for schools of gray and black cod. The captain played as if he was fishing for an Emmy. He was the third we'd had that year. The first was a Dane in clogs; then there was a raspy salt up from Seattle; now this guy, bitter, with a glass eye. Who the fuck has a glass eye these days? When we weren't catching quota, he'd soothe his anger by shooting seagulls with his twelve-gauge.

I'd been assigned to walk the deck, watching for cod that slipped off their hooks before they could be pulled aboard. I paced, fighting cold and boredom. A fish broke free and began to float down starboard side. Suddenly the loudspeaker erupted with static. "Get that fucking fish, boy. That's money floatin' by." I snatched up the thirty-foot gaff, keeping an eye on the silver and gray moving quickly down along the hull. Swells had us rocking in and out of the trough. The head's the only place to aim; anywhere else will gash the meat, and buyers refuse fish with cuts.

In one motion, I swung the gaff out past the fish and jerked it back, trying to sink the hook. I missed. Here it came: "Ahh, fuck. . . . You missed, you fucking speck of shit. Your mother went through all that pain for nothing." I ran to the stern, took another swing, missed, watched the fish disappear into the milk of the ship's wake. I walked back to my position, feeling the heat of stare. I didn't look up. From the bridge, I was the only one of his thirteen crew that he could see.

I looked down and followed the longline, forty thousand hooks coming out of the water into the belly of the boat, spaced five feet apart, squid-baited for black cod, hanging on two-foot cords attached to the hauling line. On a good day, there'd be a fish on every hook. That day, it was one for every twenty or thirty. Tommy, an artist with the gaff, hooked deep into the heads and hauled them into the boat in one smooth motion. Then he started cursing—fish were falling off the line. I reached out, the gaff so long it bent to an arch as if I was setting up for a pole vault. I pulled in hard and sharp, hand over hand. Hooked the head, pulled it on deck. Then another one, hooked perfectly. I grabbed a fish by the gills, and before I threw it into the hold, I held it up and shook it at the captain, looking for recognition. He just stared at me, silent. Took me a while to learn that quiet from the bridge meant I was doing good.

The crew I worked with was a mixed bunch. These were the early days of the immigrant race wars. Mexicans began showing up on the docks to work. They came aboard as highbrow, educated men—Juan was a dentist, Manuel and Pedro met in law school—and treated us like the white trash we were—high school dropouts, felons, drug addicts. We'd get into raging fights, a mini–class war. No one ever took a swing, but we'd get up in each other's faces, screaming full-throttle. At the time, I had no understanding of the context that brought these men to Alaska. Hell, I still don't know why they were there. All I knew was they weren't fishermen, so lots of mistakes were made. During a set of gear, Juan placed a coil of rope wrong side up, and when I grabbed it to throw it overboard, it coiled around my arm and almost dragged me with it. At sea, greenhorns will get you killed.

Now, decades later, the seasons I fished in Alaska blur together. I'm left with vignettes and bits of stories. One time, the FBI radioed our captain to report that one of the new deckhands—a lanky, twitchy plank of a man—was wanted for murder down in Eugene. The captain searched his room and found a handgun, which he confiscated, and then he locked the guy in his bunk. We were ordered to dock at Shemya Island, an eerie military base dotted with what looked like *Star Wars* domes. We had been at sea for weeks but were barred from setting foot ashore. The Coast Guard boarded and led him off the ship, and we headed back to the fishing grounds, pissed that we had lost a set of hands: more work for the rest of us.

Sometimes it's rougher on land than sea. You've been working twenty-hour shifts, with thousands of dollars burning a hole in your pocket, and you haven't seen a barstool in months. One time, we landed on one of the Aleutian Islands to resupply—I forget which one—and six of us walked into a tiny bar packed to the gills at two in the afternoon. Vic, our engineer, who was disabled from an explosion, shouted, "We're buying the bar for a day and locking the door. Stay, you get free drinks till tomorrow. If you want to leave, get the fuck out now." A few people shuffled out, but most stayed. It got ugly. I blacked out by 7:00 p.m., but not before knocking boots with a cook from another boat. Both of us took our chewing tobacco slugs out of our mouths before diving deep.

Since sleep was rarely on the menu, we fueled with calories—thousands per meal. My captain was spending fifteen thousand dollars a month on food. Our chef was an Inuit grandmother who cooked gracefully in fifteen-foot seas with the help of duct tape to keep the dishes moored to the tables—a skill known as "cooking in the ditch." A breakfast would be forty eggs for eight guys, with strips of steak, potatoes, and bread on the side. We shoveled and choked, eating with our mouths open so we could suck oxygen and swallow at the same time. We'd rarely talk and would instead watch looped videos of fishermen disappearing under the ice in botched Coast Guard rescues.

My third season fishing, there was one real foodie on the boat. He was an adventurous eater, even when measured by today's standards. Half our fish went to a Japanese company, whose massive factory ship would pull alongside our suddenly toyish 170-foot boat so we could off-load without steaming back to Dutch. To ensure quality control, they installed a man named Hiroto on our boat to cull the good cod from the bad. He spoke little English, kept to himself, and never ate with the crew. But he took a liking to me, probably because I was the youngest aboard and was able to mask any hints of weakness. As I worked the gear room, baiting hooks, he'd take a seat next to me and scoop up frozen squid to gnaw on, or belly-laugh while showing off his gut-churning talent of tossing back maggot-filled herring before lunch. Other times, he'd quietly bring me plates of fried seaweed with mayonnaise and barnacles doused in his own supply of soy sauce. His taste buds foraged far below the surface, delighting in the fullness of the ocean's bounty.

CHAPTER 4

Two Hearts, Two Minds

After several seasons of fishing in Alaska, I felt a breeze stirring. From Georges Bank to the Bering Sea, I had been on the water since age fifteen and had a good number of years of fishing under my belt. But I was no longer a kid, and the thrill of riding the cocaine rails had faded. This breeze rose slowly, beginning with thoughts of my family. I had seen little of them over the past few years, mostly just a handful of calls with my mom and my sister, none with my father.

Up until this point, I had seen myself as a full-throated hockey-playing Newf, raised in a dirt-poor province, who dropped out of high school to fish the globe. Pretty much all of my life's learnings had been soaked up on the job, and as far as I was concerned, this was a first-class education.

But now there was this unfamiliar pull. I began to feel as though I had drifted too far from the house I was raised in.

I didn't come from one class but rather from a mix of many. My sister had leapt from Newfoundland to Harvard. Similarly, my mom ascended from secretary to manager at a publishing house. And my dad started the linguistics department at the Memorial University of Newfoundland and was a renowned expert in Inuit languages. As I was growing into an adult, new identities began to push from within. I was a fisherman, yes, but I was also something else.

One morning, after we wrapped up a ten-hour shift hauling longlines, I took to the deck to sip a can of Dole's fruit cocktail I had let ferment for a few weeks up in the bow of the ship. I scanned the horizon, following the thin, dark line at the edge of the earth. I felt the urge to explore—not new oceans, but new ideas. I wanted to know why the world acted as it did, and where I fit in it. I had been raised in a home full of books, and I realized I was missing some part of myself, a part as vital as the one shaped by work. It was time to go back to school.

As I look back now at this moment in my life, I interpret it not as a sudden departure—I remained a fisherman and continued to head back to Alaska during school breaks. It wasn't some transformational moment when I concluded that commercial fishing wasn't for me. Rather, this was the moment I began to understand my duality, my selves. I carried in me two histories that I viewed as incompatible: one a history of tools, toil, and meaningful work; the other of pens, books, and ideas. It would be decades before I would find a way to live fully within both.

Picking up from where I left off in high school wasn't easy. Despite my poor track record as a son, my mom let me move into her basement in Massachusetts while I took classes to fill in the gaps. My parents were long divorced by this point. I sought advice from a guidance counselor at my old high school, who generously helped me piece together a transcript from "work experience" credits, "independent studies," and remedial classes. I applied to three schools. One was my dad and mom's alma mater, the University of Michigan, where the admissions department barely glanced at my transcript and wondered why I'd bothered to show up for a visit. The second was Hampshire College, up in Amherst, Massachusetts, considered to be a haven for alternative thinkers. I wore a white dress shirt and red-striped tie, and had practiced interview responses, hoping to impress. After being rejected and inquiring why, I was told they'd thought me too conservative for their community. Completely out of my element, I had no idea that I should have played the rebel: sported my fishing clothes, left my hair long, and thrown in some foul language. But for some reason—to this day I don't know why—the University of Vermont let me in.

Now college bound, I fished for a season in Alaska and then flew straight from Anchorage into Burlington, Vermont. Three days before, I had been in the middle of the Bering Sea; now I was on a plane, nervous and giddy, expecting to study ocean-ography and meet girls fresh out of prep school. Nope. When I arrived, I was packed into a four-story dorm infested with soon-to-be frat boys. The culture shock was extreme. For most of these kids, it was the first time they'd ever left home, and they were intent on partying. I was older and had been on my own since I was fourteen. I lasted three days in the dorm. First day, I hated the glaring whiteness of my room. Day two, I hated my roommates. Day three, I said fuck it, and moved out.

Next door to the dorm—still technically on campus—there was a golf course. In a wooded section out near the third hole, I built a lean-to, put a tarp on the floor, set up a cookstove, and moved in. It turned out that Vermont women liked guys who lived in lean-tos, so I hooked up a lot, even though I smelled like a woodchuck. I sold mushrooms and acid that I had shipped

UPS in Cheez-It boxes from a fishing friend who was living in Boulder. I didn't need the money—I was making bank on the boats—but staying illegal helped me cling to my hard-rolling fishing identity.

Classes were brutal. I tried out the oceanography major I had been fantasizing about, thinking maybe I could remake myself as the next Jacques Cousteau. But I suffered under the iron heel of math and science. I racked up F's in a bunch of classes and teetered on the edge of getting kicked out. Plus, I couldn't really see. Swabbing the deck of the ship over the summer, I had splashed some chemical solution into my eyes. It burned for days, which turned into years. I lost some sight in my right eye, and some days I couldn't look up. I'd sit in class unable to raise my head, unable to concentrate from the pain. I took a writing class and kept writing about fishing. I couldn't spell, had no concept of grammar, just wrote raw reports of the sea. My professor thought it was romantic to have a fisherman in their midst, so I suspect she inflated my grades.

At night, back in the woods, I strung hooks from gear I ordered wholesale. I'd sell and ship them off to longline fleets. Cut, tie, repeat—thousands of times. Classmates would come sit on my lean-to floor and watch, suburban kids wondering why I'd spend my free hours in menial, repetitive labor. I knew why, but I couldn't explain.

UVM at the time had a reputation as a "party school," and the eighteen-year-olds running around campus did a decent job keeping the brand alive. I sold them drugs but avoided the parties. Although I struggled in class, I was still hungry to learn, and spent many Saturday nights alone in the top floor of Bailey/Howe Library. I read randomly, often just picking up books left on tables: a religious meditation about how elephants, as lifelong monogamists, were paragons of Christian virtue; James

Joyce's letters to his wife, which could make a fisherman blush ("You had an arse full of farts that night, darling, and I fucked them out of you . . ."). I wrote a paper about the Waco, Texas, massacre of the Branch Davidian religious cult, and another on why Greenpeace's attempts to shut down the Newfoundland seal fishery were a crime against community.

Around this time, my mind started playing tricks on me again. Unlike past haunting visions, these were small and innocent. Sprinkles of daytime stars would follow me around campus. I'd feel elated, my feet barely touching the ground. Doctors would later diagnose these as epileptic auras, but in the moment, they felt like signs of a new promising life.

And, like every college kid, I had that one professor that got his hooks in me. His name was Huck, and the class focused on Walt Whitman. He read a quote in class that I remember to this day:

> This is what you shall do: . . . despise riches, give alms to every one that asks, stand up for the stupid and crazy, devote your income and labor to others, hate tyrants . . . take off your hat to nothing known or unknown or to any man or number of men, go freely with powerful unedu- cated persons . . . re-examine all you have been told at school or church or in any book, dismiss whatever insults your own soul, and your very flesh shall be a great poem and have the richest fluency . . . in every motion and joint of your body.

Guiding words from the Great American Poet. In class, my mind was stretching, but my body was yearning for the sea.

CHAPTER 5

No Fish, Our Lives

During the summer, I'd head back to Alaska. The fishing was good—too good. Now, with an adult eye, I realize I was working in a globalized industry worth billions. I had shown up on the Bering Sea at the height of the industrialization of the oceans, transformed from the small-scale fisher of my youth into a global pillager of the high seas.

As the cod came aboard, we'd toss them onto a conveyor belt, stun them with a hard blow to the head with an ax handle, slice into their bellies, and rip out the guts. Hiroto, who still played imperial judge for quality, would sort. High-grade cod—skin shimmering, speckled brown, alabaster-white flesh—were shipped off to Japan. These made up about a third of our catch. The rest were crawling with parasitic worms and lice, squirming red and white in the meat. Hiroto tossed this trash into boxes and slapped on a Grade D sticker, which, he explained with a smirk, stood for "disgusting." These were sold at cut-rate prices to McDonald's for their Filet-O-Fish sandwiches.

To feed the ravenous appetite for fast food, we ripped up entire ecosystems with our trawls. After each haul, we threw dead bycatch—basically, anything that comes up other than what you're fishing for—overboard by the thousands; our ship was surrounded by a sea of death. We fished illegally in Russian waters. There was a government-mandated environmental inspector aboard, but he spent his days shunned and threatened, cowering in a defensive crouch. A fig leaf for our plunder.

At times, it was heartbreaking: so much waste, so much death, under the grinding heel of industrial efficiency. But, at the same time, these were, and remain, some of the best days of my life. I loved my work: the solidarity that comes with working a thirty-hour shift with thirteen brothers in the belly of a boat; the humility that comes with being in forty-foot seas; the prehistoric hunter-gatherer thrill of chasing fish to the edge of the earth. I didn't know it then, but these were soon-to-vanish years—when a fisherman could still buy a house, pay for his kid's college, and, God forbid, afford good health care.

Now, as a quiet, even-keeled ocean farmer, I peer behind me and, God, how I miss being a fisherman. Yes, I pillaged the seas for McDonald's, fishing at the height of one of the most destructive forms of food production on the planet, feeding the masses with some of the most-low-quality, most unhealthy food on the planet. Yes, that Filet-O-Fish is packed with heart-stopping calories, but it was also packed with romance and meaning. For years after I abandoned the high seas, my beloved great-aunt and I made a habit of going to McDonald's for lunch. We'd order our Filet-O-Fish, find a booth to share, and take a bite. Every time, with a hint of a smile, she'd say, "Who knows, maybe it's one of yours." She's dead now, but a Filet-O-Fish from McDonald's still brings her close.

In 1992, while I was back in Alaska, fishing the summer black cod season, news broke from the Grand Banks of New-foundland. The cod stocks had crashed, and the government was closing the cod fishery—all of it. Thirty-five thousand Newfoundlanders were thrown out of work—the largest layoff in Canadian history. There was a brutality to it: thousands of boats beached, fish plants shuttered, kids banned from tossing a line off the docks, retired old-timers barred from jiggin' in their double-ender dories. Front page on the *St. John's Gazette* was Roddy Doyle, who was arrested for an act of civil disobedience because he caught one cod and ate it.

It's stunning how, under the weight of ecological collapse, a hundred years of culture can be gutted overnight. The metronome of our music and poetry isn't the ocean—it's fish. With cod gone, meaning was swept away, dignity soon replaced with anger. Fishermen are not a bitter race, but we rot from within if beached.

Predictably, a political fight broke out. On one side were the scientists, environmentalists, and bureaucrats churning out two-hundred-page chart-filled reports detailing declining stocks; on the other were the captains of industry who rejected the science and remained hell-bent on fishing the last fish. Now, decades later, the lasting effect of overfishing is well documented. Less discussed is the historic failure of environmentalists and politicians to offer alternatives for the future.

Fishermen are not a literal people. We tell stories to tell the truth, and there are two stories we tell of those years. One is of the fisherman who agrees to give up his fishing license in return for a generous government check. He buys himself a fully rigged handsome new truck. The next day, he wakes at 4:00 a.m., drives down to the docks, and begins the daily routine of staring out at the water while he drinks himself to death, wishing he were out to sea. The other tale is of a New Yorker, a woman with piles of money, who shows up in town after the cod crash, offering to turn the local fish plant into a new factory and put fishermen back to work—making seat belts for pets. As though any job is as good as any other.

We want neither money nor workaday employment. We wake before dawn, risk our lives, and destroy our bodies; in return, we are graced with a meaning so rarely found in modern life. There are certain jobs that bind people to the earth: coal mining, farming, fishing. We power and feed this country. There are no songs about hedge-fund managers or lawyers; there are hundreds about us.

I was from a younger generation and caught in the middle. Many of us believed the science that the fish were gone—it didn't take an advanced degree to see that the technology had become too efficient. Our oceans were being scoured by fleets of floating factories armed with helicopters, chasing fewer and fewer fish, farther and farther out to sea. There was no fogging the pillage.

After seeing what had happened in Newfoundland, I lost the stomach for pillage, and, like many of my generation, embarked on a search for sustainability. The course lacked shape or substance; it was more of a hazy wandering, spurred by the rejection of a bad thing, than a clear path for a good. I wasn't an environmentalist; I still wanted to kill things for a living. It was more a search for how to protect the seas so I could spend my life working on the ocean.

My last day on the Bering Sea, we were dropped off on a barren, desolate island. While we waited for the floatplane to arrive, I bathed in a waterfall that cascaded a couple of hundred feet down an overhanging cliff. I washed off the stink of fish, death, and work. This was a farewell. I hiked the cliff, walked among downed World War II Japanese fighter planes, and gazed out at a hulking two-hundred-foot fishing trawler rusting on the shoals. Wind was picking up. I said a quiet goodbye. The plane landed. We boarded. The plane lifted slowly above the water, but then began to fall. As we hit the water, the pilot yelled, "Put your head between your knees!" We did it, and he shouted, "And kiss your ass goodbye!" with a roar of laughter. He pulled on the throttle and we took off.

CHAPTER 6

Everyone's on the Hook

I had a sense of it, but didn't know the degree to which those of us working the high seas in those years were mere cogs in the grinding wheel of industrialization. We lived and died for fish. When I was on the Bering Sea, fishing had the highest mortality rate of any profession in North America—double the rate for forestry and triple the rate of miners. Being crushed at sea by machinery was the leading cause of death, with drowning a close second. This wasn't just a contemporary fate. Back in 1816, Sir Walter Scott quoted his local fishmonger as advertising, "It's no fish ye're buying: it's men's lives."

Today American-caught seafood is considered the most sustainable in the world, but the industry is at a fraction of its heyday. Hundreds of thousands of fishermen have been thrown out of work, and once thriving fishing ports have been downgraded to romantic tourist destinations. The industrial fleets pillaged too efficiently for too long, and now with so many global mouths to feed and climate change wreaking havoc, commercial fishing will never fully recover.

This sad reality wasn't caused by the fisherman next door. So who's to blame for the collapse of the fishery? Well, we can start with the global conglomerates that run 450-foot, six-thousand-horsepower trawlers with four-thousand-ton capacity and leave swaths of dead bycatch behind every haul. Blame the tickler chains and rockhoppers that enabled dragging along the ocean floor, leaving deserts of destruction. Blame bottom-feeding politicians who handed out subsidies to keep factory fleets afloat. All this shit cost serious money, and it was bankrolled by people far from the day-to-day business of hauling fish from the water. Most fishermen aren't in control of fish; 20 percent of Peru's seafood quota is owned by the second-biggest equity firm in the United States. In 2018, another Wall Street firm purchased five of the largest fishing vessels operating in Maine.

You can even blame World War II. The war required building larger and faster ships, underwater radar, and electronic echolocation, but as fighting wound down, the U.S. Navy had little use for such large, high-tech fleets. So they were handed over to the fishing industry, and what had been designed for war was now repurposed into the most advanced and efficient form of food harvesting on the planet. These massive new ships were equipped with enormous onboard freezers and processing capabilities, allowing them to stay at sea for months at a time before returning to unload their catch. Deepwater fish that had remained largely untouched were suddenly staring down the gunnels of factory trawlers.

Advances in echolocation developed during the war for locating enemy submarines were quickly taken up to locate fish with extreme accuracy. No longer do fishermen have to rely on knowledge passed down through generations to find the best fishing spots. Thanks to the war effort, they have cutting-edge devices that can pinpoint exact locations. Taking things a step further, fighter pilots and their seaplanes were repurposed to hunt bluefin tuna around the globe.

The war also created a demand for stronger and longer-lasting synthetic materials. Fishing nets became essentially un-breakable, allowing bottom trawls to run along jagged seafloors and flatten obstacles in their way, like those annoying deepwater coral reefs. Today, lost or discarded "ghost nets" can go on float-ing through the seas for decades, or even centuries, continuing to snag everything in their paths. They are held afloat by buoys and form vertical walls, sometimes hundreds of feet long, fish-ing relentlessly until they are overwhelmed by the abundance of their catch and dragged to the seafloor. There the entangled catch is devoured by crustaceans, allowing the emptied net to rise once more to the surface and continue fishing, and fish-ing, and fishing. Ghost gear in general is such a major problem that some governments, including that of France, have offered rewards for recovered nets turned in to the coast guard.

Now imagine these technologies in the modern age of glob-alization. We own more water rights than any other country on earth, but 91 percent of the seafood Americans eat comes from abroad, while one-third of our catch gets sold to other countries. It's stunning. Take our nation's prized wild salmon, for example. We currently export millions of tons of Alaskan salmon all over the globe each year, and, in return, we import millions of tons of farmed salmon. You read that right—salmon for salmon, trading wild for farmed.

The story of squid is even more baffling. According to the *Los Angeles Times*,

California squid are being caught, frozen, sent to China, unfrozen, processed, refrozen and sent back to the United

States in giant 50,000-pound shipping containers. That's right: Every year, 90% of the 230 million pounds of California squid (by far the state's largest seafood harvest) are sent on a 12,000-mile round-trip journey to processing plants in Asia and then sent back across the Pacific, sometimes to seaside restaurants situated alongside the very vessels that caught the squid in the first place.

According to Paul Greenberg, author of *American Catch,* roughly three billion pounds of U.S.-caught seafood are sold to foreign countries every year. A significant portion of those exports make the round trip to Asia and back into our ports, twice frozen. And of the fish that are caught overseas and imported to the United States, only 2 percent is given even a side glance by the FDA. Meanwhile, one in three fish caught globally never makes it to the plate, either rotting before it can be served or thrown overboard.

The question of how to fix this feels overwhelming. On an individual level, most people know by now that they should try to eat "the right fish." But what exactly is that? All the apps, the wallet cards, signaling red, yellow, and green categories for the best and worst fish to eat—they seem to change by the day, and each one will tell you something different. According to the chef and author Mark Bittman, "Buying the right kind [of seafood] seems to require an advanced degree in endangered species. Fish shopping, in short, is not for sissies, and it's fraught for anyone with an environmental conscience." It's confusing and demobilizing. It's simpler to reach for the free-range, organic chicken. No wonder that, as a nation, we eat only about fifteen pounds of seafood per year per capita. That's half the global average. We want to do better, but where do we start? That's what I was on a quest to figure out.

Starting an Underwater Garden

So you wanna start your own little ocean victory garden? First thing is getting your hands on some water. Huh? How does someone own water? Well, as ocean farmers, we don't: we lease acreage, but we don't own it. The leases permit us to grow shellfish and seaweeds, but that's it. Anyone can still use our waters to boat, fish commercially, swim—anything they want, except growing shellfish and seaweeds. So, as farmers, we own a process, not a property right.

I like this model, because it protects rather than privatizes the commons. We invite people to dive through our kelp forests and mussel reefs. We provide duck hunters and fishermen fertile new fishing and hunting grounds. Think of it as a local park and a farm in one.

To farm, you'll need between five to twenty acres. I suggest leasing ten, but plan to start small—just a few lines to get your feet wet and learn about the currents, tides, and nutrient levels on your grounds. Figure between two to four five-hundred-foot lines, which requires 2.5 acres. Later, you can scale up and use your whole lease with fifty lines.

There are three ways to lease property: from the town, state, or federal government. I shy away from federal leases, because it's a brutal permitting process. Many towns have local shellfish commissions that control local waters, and they will help you identify acreages. Otherwise, contact your state aquaculture or marine fisheries office to find a state plot.

In my state, it costs between twenty-five and fifty dollars per acre, per year, to lease property. This means my farm "land" only costs a thousand dollars a year. It's cheap as hell, and that's why young land-based farmers are coming in droves to farm the sea—at this price, it's debt free.

Every state has a different permitting process, and some are pretty complex, but GreenWave and others have been working on streamlining the process throughout the country.

Permitting can be the hardest part of ocean farming, taking anywhere from eight to fifteen months. It's often good to start by contacting the regulators to have informal conversations regarding your proposed farm plan, soliciting feedback before you make a formal application. Once your application is in, you'll run a notice in the paper. During this time, you'll want to have a round of meetings with stakeholders—harbormasters, fishermen, tour boat companies, environmental groups—to get them onboard. Initially, farmers may meet resistance. When one of our farmers met with a local harbormaster in Rhode Island, he started off saying, "There is no fucking way this is happening." By the end of the meeting, he was helping pick out sites and offering to advocate. Be prepared to answer questions about impacts on turtles and other endangered species, but these are easily answered—in more than a decade of farming along the coast, we've never had a marine mammal or sea turtle entangled. Our farms are designed to make this a minimal risk.

Once you get your hands on your lease and permit, breathe a sigh of relief, because the hardest part is over.

part **2**

CHAPTER 7

From Fisherman to Farmer

I wasn't the only one on the hunt for a better way. After the crash of cod stocks, it looked as if aquaculture was poised to become the Holy Grail for the oceans and fishermen's ills. After leaving Alaska, I decided not to return to the University of Vermont for my last year, and instead, headed back to Newfoundland.

Since the cod fishery had been shut down, word was out that the government was dumping millions into building a new aquaculture industry to replace the wild catch fishery. To get a piece of the action, I figured I needed some schooling to give me the sniff of expertise. I put on a suit and tie, trying to look the part of a serious student, and set up an appointment with the admissions office at the newly minted Center for Aquaculture and Seafood Development at Memorial University in St. John's. Almost immediately, the door slammed in my face. I hadn't managed to pass a science class since I was a kid. This cutting-edge fisheries program was for scientists and business majors, not fishermen.

Undeterred, I charted a nonacademic route into aquaculture, hustling my way into a menial job on a salmon farm. I woke that first morning feeling intense pride about my upcoming transformation: a former ocean pillager now working as a sustainable pioneer. I was part of the new trifecta: jobs for fishermen; protein for the world; an end to overfishing. As a future aquaculturist, I could continue to work on my beloved sea, and maybe one day start a fish farm of my own. This promised to be Fishing 2.0, and the chance of a lifetime for a Newf who had dropped out of both high school and college. It was looking like I wasn't at the end of the line after all.

Boy, was I ever wrong. Within days, my dreams of sustainability sank into the dark waters of factory farming. My first day of work, I was handed a shovel and ordered onto a skiff. I'd pull up to the massive floating pens, and the seas would begin to boil. Thousands of ravenous, salmonlike creatures rose to the surface, trained to gorge and shit, gorge and shit. For hours each day, I shoveled hundreds of pounds of dank-smelling pellets ground from the meat of the distant wild cousins of these imprisoned mutants. It took ten pounds of wild fish to grow one pound of farmed salmon. So, the more salmon I fed, the fewer fish swam in the sea. Once again, I had become an ocean pillager.

The mutants we grew were neither fish nor food. A few days into the new job, I grabbed our "product" by the gills and sliced one along the spine, looking for the blood-red meat of the sockeye salmon I'd fished out of the waters of Bristol Bay, Alaska. For lunch, I'd eaten them raw, plucked from gill nets, straight out of the water, still pulsating, with a squirt of lime. But no fisherman would let these new "salmon" pass his lips.

Pellet diets had stripped the color, vivid red replaced by pasty, tasteless flesh. Then there were the disease outbreaks: viruses burning sores into salmon flesh, sea lice turning up in supermarket fillets. I knew fish, and this wasn't it. We've all heard of concentrated animal-feeding operations, or CAFOs, the euphemistically named feedlots where cattle and pigs are stuffed with low-grade feed and antibiotics while they stand belly-deep in their own shit. Well, now I knew what the equivalent looked like in the ocean. And the seals—they no longer needed to hunt. It was like a drive-thru fish joint: swim up, chew a hole in the net, eat the lazy way. For the salmon, this made for a perfectly executed jailbreak: sacrifice a few fellow prisoners to seals on the way out, but benefit the thousands who could escape into the wild.

Two days a week, I had a side gig working at the Logy Bay Marine Lab, a place I used to visit as a kid. The wooden structure was shaped like a spaceship, with round disks connected by glass hallways. It was otherworldly, perched on a cliff overlooking the sea, with golden grasses, wildflowers, and blueberries in the summer. Seals were stored year-round in large tanks outside; as a kid, I'd take buckets of bait and feed them on weekends.

More than a decade later, I was back, knocking at the lab door. I got a job sitting in a chair, squinting into a microscope, and counting rotifers. Rotifers are little creatures used in salmon feed. I spent endless hours just counting—my worst nightmare, having struggled in even remedial math. I'd lose my count and have to start again and again. So I started making up numbers and sneaking away to feed the seals.

One lesson stuck with me from the Logy lab. A room on the ground floor was dedicated to growing halibut, a most delicious and powerful fish. I'd manhandled my share in the wild as they came aboard. Some were six feet long and an arm's length across, with strength enough to knock me on my ass. But the ones in the tank were way smaller and swam in dizzy circles—and they kept dying. The lab techs couldn't figure out why. Turned out it was the door slamming shut every time technicians came and went. The juvenile fish were sensitive to vibrations, and died from shock. So they put in a door silencer, and the halibut lived. At least, that was the working theory. The lesson I learned from that? It's really hard to grow fish.

At night, I started taking classes at Memorial University in St. John's, trying to pocket my college degree. UVM would grant my degree if I retook all those freshman year classes I had failed. There was a student-run bar called the Breezeway in the basement of the college. I downed pints while scribbling cheat sheets to tuck into my sleeves during exams. Dinner was equally depressing: fish and chips made from cod shipped in from Russia.

Late at night, after work and school, I'd head back to the bar and double-fist coffee and Guinness, digging through the history and future of aquaculture. I pored over exposés from the Natural Resources Defense Council, and then, in the morning, it was back to the grind, fattening up fish that created more problems than they solved. Turned out my day job was a microcosm of the dark industrial future of ocean agriculture.

CHAPTER 8

Sea of Sunken Dreams

S hould we be farming fish in the ocean? Raise the issue within the aquaculture industry and they go fucking nuts. Hate mail piles up, and you're guaranteed to be shunned at conferences. I get it; the aquaculturists are under weekly assault by conservationists, often unfairly, and in the era of climate change, the world is going to need a shitload more food to be grown out at sea, like it or not. While others turn a blind eye, the aquaculture industry shoulders the yeoman's job of searching for a path forward.

But stand with me at the edge of a floating salmon pen and see the thousands of ravenous fish, the murky water roiling with scales and fins, and the simplest of questions will haunt you: Does it make sense to grow creatures that want to swim away?

Take a trip to another farm, this time on my boat, in the Thimble Islands. Peer over the gunnel at my saltwater fields of kelp and shellfish, and witness a few of the thousands of ocean crops that neither swim nor need to be human-fed. A distinction simple but profound. No pens, no feed, no nets—overhead costs that add up quickly—and no fish.

So apologies to my fellow fish farmers, but the foundational question of whether to grow fish in the ocean needs asking. Since the birth of ocean farming, humans have assumed that fish were the crop of choice for aquaculture. Makes sense: wild fish have fed millions, so let's grow what people already like to eat. But should we? I remain on the fence, unable to settle on a final answer.

Fish farming started in China sometime between 2000 and 1000 B.C. with freshwater carp. The first detailed description appeared in 475 B.C., in the writings of Fan Li, who described "aqua husbandry" in both mystical and economic terms in *Chinese Fish Culture Classic:*

> King Wei of Chi, upon learning that Fan Li was visiting in neighboring Lau, invited him over and asked . . . : "You live in a very expensive house, and you have accumulated millions. What is the secret?" Whereupon Fan Li responded: "Here are five ways of making a living, the foremost of which is in aquatic husbandry, by which I mean fish culture. You construct a pond out of six mou of land. In the pond you build nine islands. Place into the pond plenty of aquatic plants that are folded over several times. Then collect twenty gravid carp that are three chih in length. . . . During the fourth moon, introduce into the pond one turtle, during the sixth moon, two turtles: during the eighth moon, three turtles. The turtles are heavenly guards, guarding against the invasion of flying predators. When the fish swim round and round the nine islands

without finding the end, they would feel as if they are in natural rivers and lakes. . . . The total harvest can render a cash value of 1,250,000 coins. . . . The following year, the take will amount to 5,150,000 coins. In one more year, the increase in income is countless.

I might try Li's turtle trick to keep sea ducks from using my farm as an all-you-can-eat shellfish buffet.

Since Li grew food underwater, should we consider it farming? Historians say yes, but I don't buy it. Li's operation required gathering pregnant carp from the wild and moving them into man-made ponds. To me, a system that requires heading out each season to gather new fish falls more into the job category of hunting-husbandry than farming. What historians call aquaculture was more just fattening wild fish in reservoirs.

Compare it with land-based agriculture. Pigs were first domesticated around 13,000 B.C., which meant humans had to figure out how to do two things they had never done before: breed livestock in captivity, and keep them alive long enough to eat them. The newly minted pig farmers didn't organize raids into the bush to steal piglets from feral sows each year—too time-consuming and risky. They bred and fed their pigs. So Li had half of the equation figured out—keeping carp alive in artificial ponds—but never cracked the code of reproduction. It took a few thousand more years for humans to figure out how to breed fish for food in a controlled way.

Li's methods flourished for hundreds of years but then, suddenly, in A.D. 618, the industry was shut down by the ultimate permitting authority: the emperor. As any ocean farmer knows, it's hard to keep regulators happy. In a stroke of bad luck, the new emperor was named Li, which also happened to be the word for "carp." You'd think this would be a marketing bonanza, but

Emperor Li decided he didn't like having his name used for an edible bottom-feeder, so he issued an edict that any mention of raising, killing, or eating carp would be deemed an insult to the throne and a criminal offense. Rather than rotting in jail, many carp growers threw in the towel or started raising goldfish for ornamental purposes.

But hurdles can breed innovation. The emperor's ban sparked a revolution in fish culture that still has resonance today. A subset of growers figured out they could grow varieties of carp that weren't associated with the emperor's name. Silver, bighead, grass, and mud carp began hitting the menus. And it turned out that growing a mix of fish together was more productive than Li's carp mono-cropping. Different species complemented one another by eating different types of food and preferring to cluster in different strata of the water column, which meant that four types of fish could be grown together in the same farm. This marked the birth of underwater polyculture—an innovation later abandoned by twentieth-century industrial salmon farms, but the root inspiration for 3D ocean farming.

Despite regulatory hurdles, "aquatic husbandry" continued to slowly spread—most likely evolving in tandem around the globe. It remained simple in form, essentially just fattening fish in ponds, and the history is sparse and quirky. In 300 B.C., the Indian philosopher Kautilya drafted instructions for quickly poisoning carp farms when invading armies approached. The last thing you want is to provide your enemy with a feast of fattened fish when they land on your shores. The Romans, always on trend, built fishponds in their houses as a status symbol, serving *very* fresh seafood to guests. By the Middle Ages, it became common for monasteries to maintain fishponds, and they emerged as hotbeds of innovation, growing new species such as eel and pike, designing complex flow systems. And as far away as Bolivia, snails and carp were grown in intricate pond systems starting in the thirteenth century.

Oppression also bred innovation. In Java, around the fourteenth century, Hindu rulers exiled conquered peoples to distant coastal regions and neighboring islands, and barred them from practicing agriculture, wearing clothes, or building fishing boats. These outcasts began constructing the first fishponds, to grow milkfish, mullet, and shrimp, which fed on the blue and green algae that grew in the bottom of the ponds. Similar saltwater ponds appeared in Hawaii, where islanders built stone and coral walls over two hundred feet long on top of reefs, called *loko umeki*, to trap and grow jacks, barracudas, and mullet.

In fourteenth-century Europe, budding engineers were building their own complex ponds and canals, spurred by the royal edict of Emperor Charles IV advocating widespread fish culture. In what is now the Czech Republic, fishponds covered more than 185,000 acres. Much later, Štěpánek Netolický, probably the first-ever fish-farming consultant, traveled throughout Europe and became a celebrity for his prowess in farm design.

These decades birthed the first successful experiment in breeding finfish in captivity. This allowed for a wave of intensified fish culture, and the birth of fish farming as we know it today. A seminal fish-breeding treatise published in 1547 by Ioannes Dubravius, a contemporary of Netolický, went through four printings and was translated into many languages. Soon the printing presses began churning out books popularizing aquaculture. *Cheape and Good Husbandry* and *Certain Experiments Concerning Fish and Fruits* climbed up the charts. Seen as the father of aquaculture, Dubravius was later honored with a headstone engraved with a fish.

But then, following in the footsteps of Chinese Emperor Li, Henry VIII's Protestant Reformation killed off the budding industry. In the spreading religious revolution, targeting Catholic lands and infrastructure, fishponds fell victim as collateral damage. Monasteries, those hotbeds of fish-farming innovation, were destroyed by the hundreds. Bizarrely, seafood as a whole was used as a tool of repression; Henry even imposed fines on anyone busted buying fish from foreigners, so that he could keep hammering the Catholics and their seafood-heavy religious feasts.

By 1883, aquaculture burst back onto the world stage. The timing made sense. Industrial pollution was killing off wild stocks. Dams and irrigation canals were multiplying to support land-based agriculture, blocking migratory paths for fish. The International Fisheries Exhibition in London, which drew representatives from more than thirty-one nations, promoted aquaculture as the wave of the future and captured the popular imagination. Far-flung visitors, hailing from New Zealand, South Africa, and North America, rushed back to their home countries to convince their governments to invest in fish farming. Eager scientists stuffed millions of salmon and trout eggs into their luggage and journeyed home to build fish hatcheries.

Slowly but steadily, aquaculture emerged as the release valve for ecologically unsustainable industrialization. Fisheries, scientists, and policy makers began to see aquaculture as a way to relieve pressures on wild fish stocks and the negative effects of industrialization. For example, when salmon populations were threatened in nineteenth-century Europe, there was a huge push to grow salmon. When leaders of the U.S. logging industry came under fire for degrading salmon spawning grounds, they pointed to fish hatcheries and farms as a way to "fix" the problem without having to modify their destructive harvesting practices. When it was discovered that cod and tuna stocks were dwindling, gobs of funding were funneled into failed efforts to grow cod and tuna. Aquaculture was the quick fix to these bothersome environmental problems, and forever became too closely yoked to the wild fishery.

This was to become the Achilles' heel of aquaculture. Time and again, we see an industry forced to scale up to replace fish stocks decimated by overfishing and pollution. Too many people look to aquaculture as the way to maintain "business as usual." If the fish are dying, just grow more. Today in China, there are entire islands dedicated to growing billions of fish that are released into the South China Sea so that fishermen have something to catch.

By 1914, the excitement began to die down, as farmers and investors began to realize that growing fish was complicated and expensive. Most hatcheries and fish-farming operations in the United States shuttered. This was partly due to a lack of understanding about reproductive biology, nutrition, and disease, but also because fish are capital-intensive, requiring winterized ponds, egg incubators, and mechanical saltwater systems. Critics began questioning whether the investment was worth it, faced with the lack of strong evidence that hatcheries were making a significant impact on rebuilding fish stocks. By the time government resources shifted into ramping up for World War I, most farming operations had been closed or converted into research laboratories.

THE BLUE REVOLUTION

Over the next decades, the post–World War II industrialization of the wild fisheries flooded global fish markets, creating little incentive to farm what could now be caught at a previously unimaginable scale. But by the 1970s, harvests began to level out, because of both overfishing and newly established exclusive economic zones, which barred global fleets from fishing within two hundred miles of another nation's shores.

Falling catch rates once again opened the door for aquaculture. Between 1970 and 1989, total farmed production increased by 40 percent. Today, the yearly aquaculture growth rate is three times higher than the rate at which terrestrial agriculture grew over the past several centuries. By 2013, fish farm production topped beef production. In 2019, we're eating more farmed fish than wild caught; around 70 percent of the salmon we eat comes from farms. It's big money. Globally, the aquaculture industry is worth almost a hundred billion dollars.

But, as pretty much every consumer knows, these growth rates came at high environmental cost, a fact that haunts the industry to this day. In 2011, the *New York Times* editorial board summed up public perception pretty well:

> Aquaculture has repeated too many of the mistakes of industrial farming—including the shrinking of genetic diversity, a disregard for conservation, and the global spread of intensive farming methods before their consequences are completely understood.

Just as land-based agriculture began searching for more sustainable ways to feed the planet, aquaculture accepted the trade-off of high yield at the expense of the ocean's health. This meant pumping carotenoid and egg yolk additives into the meat, using antibiotics to keep diseases at bay, feeding consumers pesticides for dinner, and dumping chemicals such as emamectin benzoate into the salmon feed. Farmed salmon became notorious for the toxic PCBs that showed up in tests.

Another major problem, which continues to this day, is the mass escape of farmed fish into the wild. Chile, for example, has never been home to wild salmon, but its coastlines and rivers are now populated by invasive salmon that escaped from farms. This salmon "leakage" is constant. The Atlantic Salmon Federation estimated that more than two million salmon escape into the North Atlantic each year. Why is this a problem? Farmed salmon are known to be more aggressive than wild fish, and they're way bigger, since they have been bred to grow as large and as fast as possible. These traits allow them to dominate the Atlantic wild salmon for food and mating privileges. With the ever-present risk of salmon escape, it's not surprising that Alaska has banned fish farming to protect its wild salmon fisheries.

Disease still runs rampant. According to Liesbeth van der Meer, Oceana's representative in Chile, who is a world expert in salmon aquaculture, the deadly bacterial illness piscirickett-siosis is still common on farms, a disease that passes to wild fish who swim past the farms. Chemicals then used to treat the disease—along with feces and uneaten food—build up a layer of toxic gunk on the seabed beneath the farm, only worsening the unsanitary conditions. Untreated salmon waste pollutes waterways, with an average-sized salmon farm, containing 3.5 million fish, producing as much daily waste as a city containing 169,000 people.

And, of course, much of the industry remains hooked on growing carnivorous species. This requires ever-increasing volumes of fishfeed, specifically in the form of fishmeal and fish oil from wild-caught anchovies, sardines, and other small species. So, in the end, fish farming causes more fish to be taken out of the ocean, not fewer. China, for example, diverts more than half of its wild-caught fish for aquaculture feed, and still must import over seventy-five million pounds of fishmeal from the United States each year to meet demand. A large percentage of that imported fishmeal comes from menhaden, a keystone prey species for wild fish in the States, which means less food for our wild fish to eat.

The result: aquaculture has ended up with one of the worst brand names in the grocery aisle. As Mark Bittman, the taste-maker of ethical food, said: "I'd rather eat wild cod once a month and sardines once a week than farm-raised salmon, ever." Michael Pollan declared to *National Geographic,* in their "Future of Food" series, that aquaculture "fundamentally is not sustainable." At the state level, opposition from commercial fishermen and the public at large continues to grow. In March 2018, after three decades of experience with aquaculture, Washington State banned all Atlantic salmon farms.

Once consumers caught on and began boycotting farmed fish, the aquaculture industry adopted a furtive strategy of mis-labeling their seafood and doubled their marketing budgets. Fish were secretly shipped off to processing facilities that mixed farmed and wild. Thousands of untraceable frozen fillets were trucked off to restaurants and seafood markets. By the time the poached salmon hit the dinner plate, it was seasoned with lies. To this day, one out of three fish sold in the United States is not what it promises to be, and, globally, one in five fish is mislabeled.

Mother Nature abhors monoculture. Stuff any animals cheek to jowl into overly crowded pens and the oceans will fight back with rampant disease. It's pretty simple: too many fish shitting in one place create fertile swamps for disease and pests, which in turn trigger a cascade of antibiotics, pesticides, and GMOs. It's a death match. This shouldn't come as a surprise: replicating industrial animal farms at sea comes with the same bucket of problems. As is the case on land, the oceans thrive on diversity.

Those blurry Newfoundland nights at the bar, delving into all this history of aquaculture, were eye-opening. Yes, the aqua-culture world I was working in day to day had major problems, but, although I didn't know it then, the underlying premise of transforming fishermen into farmers marked the beginning of a journey that would consume the next twenty years of my life. The question of how to farm the sea had captured my head and heart.

EVEN THE WORST IS BETTER

Over the years, a sector of the fish-farming industry split off and went on its own search for sustainability. Inevitably, some companies were insincere, marketing sustainability without actually doing it—classic greenwashing. But others have made significant and meaningful advances.

Some have gone on a journey to find the right fish to farm. Companies like Australis Aquaculture have settled on barramundi, or Asian sea bass, which are naturally adapted to crowded environments, making them well suited for aquaculture. They are also passive when handled by humans, fertile year-round, and naturally disease-resistant. Another favorite is arctic char, which is from the same family as salmon, has a low feed conversion ratio, and, like barramundi, can handle high stocking densities. Most are grown in recirculating, aboveground tanks in Iceland and Canada.

Strides are being made in the feed sector as well. Some farming operations have been able to drive down their wild fish feeding ratios from six pounds of wild fish per pound of farmed fish to nearly one-to-one. Investment is pouring into nonwild fish feeds. In a partnership between the National Oceanic and Atmospheric Administration (NOAA) and Kampachi Farms in Hawaii, the 0-percent fishmeal feed was tested on farmed kampachi, with excellent growth rates.

Some are even searching for seaweeds as an alternative. Fish farmers have finally realized they don't have to feed their fish wild fish to get the high levels of omega-3 fatty acids that consumers are demanding. Fish don't make omega-3's themselves; they can get them from eating seaweeds. The Ireland-based company Ocean Harvest Technology has developed a seaweed-based salmon feed alternative called OceanFeed that has been shown to produce salmon with higher levels of omega-3's than those grown with a standard diet. Adding seaweed to the feed also helps to boost the immune system of the fish, reducing a farmer's reliance on chemicals and antibiotics.

Others are using shellfish and seaweed to soak up the pollution from salmon farms. They've branded this practice as "integrated multi-trophic aquaculture" (IMTA). According to Thierry Chopin, an aquaculture expert at the University of New Brunswick and one of the founders of IMTA, "Farmers combine the cultivation of fed species such as finfish or shrimp with extractive species . . . shellfish and other invertebrates that recapture organic particulate nutrients for their growth."

Slowly but surely, these farmed fish have been making their way onto the Monterey Bay Aquarium's list of "Best Choice" seafood, which is seen as the gold standard for seafood rating. Australis products are now sold in huge grocery store chains, including Stop & Shop and Giant, and have also appeared on the menu of the three-Michelin-starred restaurant The French Laundry.

According to seafood hero Barton Seaver, a chef, author, and director of the Harvard School of Public Health's Sustainable Seafood and Health Initiative, "As a chef who once quite vociferously preached that aquaculture across the board was 'farmed and dangerous,' I don't regret the passions that drove me to that position, but I do proudly sing a redemption song." I'm not quite as bullish as Seaver, but I'm sympathetic.

The less talked about finger on the scale in favor of aquaculture is that, in terms of climate change, even the worst aquaculture is better than any land-based livestock farming. Farmed salmon has such a poor reputation that when consumers head to the grocery store they end up walking past the seafood counter to pick up free-range, organic chicken for dinner. What they don't know is that even the big, bad, and ugly fish farms have a lighter environmental toll than the producers of beef, pork, or chicken.

If we compare farmed seafood with terrestrial proteins, measuring each by the environmental impacts, greenhouse gas emissions, antibiotic use, freshwater use, and feed conversion ratios, even the worst farmed seafood comes out ahead. NOAA lists feed conversion ratios for beef, pork, and chicken as 8.7, 5.9, and 1.9, respectively. Compared with their estimate of 1.2 pounds of feed to produce one pound of salmon, the salmon looks like a pretty good alternative.

A report by the Environmental Working Group in 2011 found that farmed salmon production creates less than half the greenhouse gas emissions as beef, with most coming from the production of fishmeal. That's still a lot of greenhouse gas, but it's way better than land-based livestock.

So, if you're a climate change warrior, you have an obligation to get your food from the sea. We have moved beyond the luxury of whether to farm the oceans. The only question is what model to deploy.

ASK THE OCEAN WHAT TO GROW

Our global food system is in crisis. According to a 2017 UN report, the global population is expected to reach 9.8 billion by 2050. Coupled with rising consumption of meat and biofuels, food production will need to increase by 70 percent by 2050 to meet projected demand.

Unfortunately, land-based agriculture is not poised to double in production anytime soon. A peer-reviewed study in 2013 determined that the yields of four of the world's key staple crops—maize, rice, wheat, and soybeans—are increasing at rates far slower than projected demand. In fact, according to the agricultural expert Jonathan Foley in an article in *The Washington Post*, farmers in many parts of the world are literally "starting to hit a 'biological wall,' a limit on how much yields can keep rising."

Climatic stress is a major culprit for many of the challenges facing agriculture, and as maximum temperatures continue to increase, we are going to see widespread crop failures and serious reductions in yield. A recent study in *Nature Communications* found that maize, a crop that underpins so much of our food system, is very sensitive to extreme temperatures, putting a constraint on its future production. Water—be it too much or too little—is of similar concern. Terrestrial agriculture accounts for 70 percent of freshwater usage, and according to the World Resources Institute, more than a quarter of this farming occurs in areas of high water stress. In 2012, about 80 percent of agricultural land was affected by drought, and in January 2013, the USDA designated 71 percent of the United States as a drought disaster area. By as soon as 2025, 64 percent of the world's population is expected to be struggling with water scarcity.

California, which accounts for a significant portion of American agricultural production, is in the midst of a record-breaking drought. This dry spell is only a taste of what is to come, according to a recently published NASA study. Scientists predict a bleak outcome for the American Southwest and Central Plains, with an 80 percent chance of a megadrought lasting thirty-five years, between 2050 and 2099, involving a sustained period of low rainfall, snowfall, and soil moisture. The drought is forecast to be the worst in a thousand years.

On top of climatic stress, working farmland is in constant competition with urban development as our population continues to grow. According to the American Farmland Trust, more than an acre of farmland is lost to development every minute. In just the past twenty-five years, over twenty-three million acres of land were lost, corresponding to an area the size of Indiana. One result is that prices for prime farmland are skyrocketing. To make matters worse, according to Mark Bittman in his 2009 book, *Food Matters,* "It takes 70 percent of all of the available farmland in the world to produce the meat we're eating now." It takes two thousand gallons of water to produce one pound of chicken. In short, he says, we simply cannot continue farming as usual, because "the land just isn't there."

With increasing pressure on the land, we will need to turn to the ocean to provide our food. Ocean foods are healthier, more productive, more efficient, and less consumptive of resources than land-based agricultural foods. But there just aren't enough wild fish to fill the need. More than 2.6 billion people already rely on the ocean for their main source of protein, and two-thirds of the world's largest cities are near the coast, with rapidly rising populations as people flock to urban areas. Consumption of fish around the world doubled between 1973 and 2007, and, according to the UN Food and Agriculture Organization, we have probably hit the limit of wild-capture fisheries; we are eating more seafood, but we can't rely on wild stocks to meet the demand.

The ocean is the next big frontier for food production, and we have an exciting opportunity to do it the right way. However, much of the aquaculture industry continues to march down the path of growth-at-all-costs and to wreak havoc on our oceans. People living and fishing near the large farms are still pissed. In April 2018, fifty British Canadian chefs joined the Vancouver Island First Nation to force the closure of twenty salmon farms in the region. Farther north, in Alaska, a state that has banned all finfish farming, commercial fishermen showed up in force during a pro-aquaculture meeting organized by the Trump administration. In August 2018, escaped salmon began turning up in the nets of Newfoundland fishermen. It turned out that Cooke Aquaculture, one of the largest salmon-farming companies in the world, had failed to report escapes from one of their farms. Twelve months earlier, on the Pacific coast, Cooke suffered another salmon breach, which Kurt Beardslee, of Wild Fish Conservancy Northwest, deemed "an environmental nightmare."

Despite popular concerns, billions upon billions of dollars are being thrown into farming fish. Chile-based Ocean Arks Technology is raising money for a 550-foot-long offshore robot aquaculture "Ark" supposedly capable of producing four thousand metric tons of any commercial fish species. One Ark costs twenty million dollars. In China, a monster they are calling an "offshore super fish farm" is in the works and expected to cost a whopping billion dollars per farm.

And of course there is the threat of so-called Frankenfish. The Massachusetts-based company AquaBounty has genetically modified Atlantic salmon with antifreeze proteins, so fish could be grown in colder water, and with growth hormones taken from ocean pouts (they look like eels) and other salmon species. AquaBounty's fish are available in Canadian grocery stores, although the five tons sold in 2016 were not labeled, and the company refused to disclose the names of purchasers. Here in the States, Alaska's Senator Lisa Murkowski has led the fight to block all GMO seafood, including the introduction of the Genetically Engineered Salmon Labeling Act in July 2017.

Escapes, disease, feed conversions, opposition from commercial fishermen, massive financial outlays, GMOs—sounds exhausting. Maybe my thinking on the problem is too rudimentary, but as a non-fish ocean farmer, to this day I can't get past the simple question: Why grow fish if there are thousands of potential crops that don't swim, are cheap to grow, and feed naturally on what's already in the water?

I suspect this question is rarely asked, because the founding operating principle of the modern aquaculture industry was to grow based on existing market demand—farming what Americans already wanted to eat. The problem is that those consumer preferences are based on the wild fishery. We're growing what people have eaten in the past. Just because we've successfully hunted salmon and tuna in the wild for hundreds of years doesn't necessarily mean that those are the species we should grow. For too long, aquaculture has been yoked to the wild fishery.

Instead, we need to approach the ocean as a unique agriculture space, and ask: What do you want us to grow?

CHAPTER 9

Landlocked and Drowning

With both commercial fishing and industrial aquaculture crossed off my list of professions, I begrudgingly looked to the land for work. I called up my former UVM Walt Whitman professor to get some career coaching. He'd always been able to inspire, and I knew he wasn't enamored with the ocean-faring life. When he was teaching *Moby Dick* in class, I remembered him offhandedly saying that being at sea for months at a time sounded monotonous. So I figured he'd be the right person to ask what the hell I should do on land.

On the side, he moonlighted as the speechwriter for a local Vermont politician and offered to get me a job. Looking back, I don't know why I said yes. Sure, my dad was a refusenik during the Vietnam War, but I knew nothing about politics. Never voted, never cared. Even though my experiences working on the ocean had planted seeds of working-class environmentalism, that was about fishing, not elections.

I took the job anyway. Over the next few years, I tossed around doing political work and community organizing, learning a whole new set of skills, working to engage dairy farmers, veterans, even coal miners. Then someone I respected encouraged me to go to law school. This was even further outside my comfort zone. I had never done well in school, and was told that I likely had a slew of learning disabilities. I wasn't a dunce, but everything I knew was self-taught and hands-on. But she insisted I could do it, and even promised to help me get into a good school. I felt the sea had forsaken me, and the prospect of being an educated man was a strong push and pull. The following year, on the strength of her recommendation alone, I started classes at Cornell Law School.

Getting into Cornell was the high point. Then it sucked. From day one, all my insecurities about education and whether I was "smart" were inflamed. The books were often five or six hundred pages long, the font minuscule, with footnotes by the thousands. I began waking at 5:00 a.m., back to my fisherman's hours, to read until the pages blurred. Although miserable, I did well in one class—first-year contracts—emerging with an A+. I had called a friend from college to ask how to study for tests, and he said to write and memorize a sentence in which each word represented some aspect of contract law. I defaulted to pornographic sentences and aced the test, lewd images dancing in my head.

That A+ gave me a boost of confidence, but it turned out to be a fluke. I eked by in the rest of my classes, often skipping weeks of lectures just to avoid getting called on by professors. I felt stupid and alone, and compensated by trying desperately to keep a finger hold on my identity by hiding a spittoon in my inside jacket pocket so I could chew tobacco during class. I tried, but I just fucking didn't belong there. Grasping for a lifeline, I joined a pickup hockey league. Most of my team worked the snowplows for the Department of Transportation and kept a firehose of Michelob Ultra on tap for before and after games. It had been a long time since I took to the ice—my Canadian instinct had long disappeared—so I sucked, but the chat and funk of the locker room was enough protein to keep me going to classes.

When I was getting credits for legal clinic—a program meant to give hands-on legal experience to students—I was assigned to help represent a janitor who was barred from collecting unemployment because he had been fired for sexual harassment. It was a weird case, because he was canned for telling a nurse that her puppy "smelled like his mom." The night before I was scheduled to present at his unemployment hearing, I played a hockey game, and got hit from two sides at once. My helmet cracked, and I woke up on the ice with yet another concussion. Next morning, when I showed up at the hearing, bandage on my forehead, my vision was still blurred, and I couldn't read the notes I'd prepared. My client shook his head, rightly worried. A better-suited student, who, unlike me, was destined to become an actual lawyer, took over.

I made it to graduation, but never took the bar. Why did I even try? In retrospect, when I look back at this period of my life, I feel the spike of shame. Clearly, it was a mistake—an honor to be accepted, but a waste of a few good years of my life. And, Jesus, even with the scholarships, it took me over a decade to pay off the debt.

If I'm honest with myself, I headed down the wrong track because of an internal war. I was proud of my blue collar knowledge but was considered by others to be an "uneducated" man. At the time, I didn't see any way to make a life on the water, but I was just as lost on land. And, above all things, my father valued education; even though it had been a decade since we last spoke, part of me still wanted to make him proud.

Years later, after I was well known as an ocean farmer, I would often get asked the question: How did someone like you create all of this? The first time, I answered by explaining my full history, including law school. As soon as people heard about my "educated" detour, they'd say, "Ah, that's how you did it. Now it all makes sense."

This pissed me off, because it was an insult to my lived experience. The real source of my learning is the "professors" of blue collar innovation. As fishermen, we're marine biologists, field engineers, inventors, and entrepreneurs. Just because one of us takes a wrong turn and studies law doesn't mean it's the source of their success. I became successful as an ocean farmer despite law school, not because of it.

Out of frustration, I ended up dropping Cornell from my story, because it ran so counter to my life's work. My greatest influence is not the Ivy League; it's a lifetime on the water.

OFF THE GRID

After fleeing law school, I was still landlocked and didn't have a plan or place to go, so I followed my heart and tried to build a life with my on-and-off-again girlfriend. Unfortunately, she lived in the one place worse than law school: the suburbs.

First day I showed up in town, a police officer approached my girlfriend in the Dunkin' Donuts parking lot while I was getting coffee and asked, "Are you all right, ma'am?"—investigating his hunch that I was an abuser. In a town with no black folks, I guess police defaulted to harassing down-market whites.

I picked up a seven-hundred-dollar 1961 Airstream trailer from a haggard hippie in Vermont, parked it in the woods across from my girlfriend's parents' house, and began living off the grid. I humped water, shat in a port-a-potty, ran an extension cord through the woods to get power. I signed up for a $9.99-a-month gym membership at Planet Fitness so I could shower once a week. Couple hours a day, I parked in front of a coffee shop to steal the internet.

Friends of a certain bent thought I was living the dream. During the first few months, so did I. But as years went by, my Airstream romance devolved into filthy hard living. In the winter, ice would form on my cups of water; in summer, the metal shell became a prison of sweltering discomfort. My girlfriend spent more and more time at her parents' house. My plan was to spend a year in the woods, make some money, then find a house. One year became seven.

An emptiness swallowed my days. Sluggish hours of daylight, evenings floating in shots of Ballantine's and pints of Guinness. I drove a lumber truck, delivering reclaimed wood into Manhattan, while searching the shoreline for a new fishing hole, but pulled up nothing.

And my head was hurting. One summer afternoon, I headed into New York City to wander around the upscale neighborhoods of SoHo. It was hot; crowds pushed along the sidewalks. I saw a flash, a wave of terror flooded over me, then blackness. The paramedics found me flopping like a flounder at the corner of Bleecker and Thompson Streets, grinding my face into the pavement. I woke in a puddle of blood and confusion in the emergency room. My face was pulp. A nurse told me I'd had a grand mal seizure.

I spent the next three weeks in the hospital. It was relatively rare for adults to suffer their first seizure in their twenties, so the doctors assumed it was my five concussions, tallied up through years of hockey, bar brawls, and fishing accidents, that were triggering the fits. But brain scans showed no tissue scarring, so head trauma was crossed off the list.

Next they tried to induce seizures to map the source, by wiring me up to an EEG machine, putting me on round-the-clock video surveillance, and prescribing three days of hedonism. Nurses set up strobe lights, plied me with alcohol and a river of coffee. I was forced to stay awake for over twenty-four hours straight. Still no seizures.

My bunkmate, a Teamster who worked at the Javits Center, had convulsion after convulsion behind his curtain. I told the nurses that coffee, no sleep, and a few drinks weren't stressing my system—that combination was a fisherman's daily bread. But I did learn that my auras originate in the part of the brain that generates fear. My neurons become overstimulated, over-electrified, and I am hit with extreme waves of terror. Intense, wild terror, escalating to the highest levels that humans experience. I feel the fear of being attacked by lions, freefalling during a plane crash, drowning. It was the short straw: other epileptics experience visions of God, bouts of hypersexuality, or even hypergraphia—intense, prolonged desires to write. I just get scared shitless.

Unable to induce a seizure, the doctors asked me to plumb my history for clues. It was a hard task, because I didn't know what I was looking for. Had I ever seen or heard or smelled anything that wasn't there? Had I ever lost periods of time? I listed any weird shit I could remember: hallucinations of man-sized bugs when I was nine; the times in college when colors would melt together, followed by euphoria, then waves of confusion; that weird man's face that haunted me whenever I shut my eyes in Alaska. And, of course, the long string of PCP and acid trips in my teens.

As I told them what I could remember, the doctors nodded knowingly, but for me none of this felt too far out of the ordinary—everyone's brain plays tricks on them. I was left with a strange, haunting foreignness—a feeling I haven't yet shaken, decades later. My mind is not fully mine. Under the constant threat of seizures, the line between my brain and myself is hard to slice, so the question of what's the "me" in me slips in and out of focus.

Released with a script for meds, I returned to trailer life. Although the full-blown seizures were blocked by evil little six-hundred-dollar-a-month pills, which were mailed to me by the pharmaceutical company as part of their low-income access program, the pre-seizure auras continued.

My meds slowed my mind to a crawl. I had to prepare written notes before I spoke to people, because I kept losing my train of thought. I struggled for purpose, unable to keep sea dreams alive. I read that in the Czech Republic they fried carp and potato salad as the traditional Christmas Eve dinner. Even that free fish were handed out to garner votes in support of the referendum on joining NATO. Surely, the time was ripe for suburbanites to embrace Czech cuisine. I hatched a limp—and, in retrospect, quite embarrassing—plan to empower Stepford wives to grow carp for dinner.

I bought a plastic tote from Walmart, an aerator, and one juvenile carp, and set them up in the trailer. I was confident my trailer experiment would emerge as the new, updated World War II victory garden—turning every house into a DIY fish farm. Keeping the bills low, I ate beans, rice, and fruit. So the carp ate beans, rice, and fruit. After a few weeks, he threw himself out of the tub and dried to the floor. I scraped him up with a spatula and threw him back in the water. For his remaining eleven days, he could only swim backward. When he finally gave up and floated to the surface, I tossed all four inches of him into the woods. A truly unsustainable death.

I needed to make money—preferably under the table—so on weekends my girlfriend and I began humping the two hours into New York to hustle what I thought was going to be pocket change from tourists. She sold hand-silk-screened shirts. My big sellers were "reclaimed words on reclaimed wood." We were both working at the lumberyard that sold reclaimed wood, and at night I'd pick up the scraps. Each chunk had a story: Coney Island *ipe* from the boardwalk; redwood from Brooklyn water towers; vermouth tanks from a Queens distillery; oak from an old porno theater that had been demolished in Times Square.

I scoured old dictionaries and word-o-phile websites for words that had fallen out of usage and captured something ephemeral: *petrichor*, the smell of rain on dry earth; *limerence*, the first moments of love; *ruckle*, the last shuddering breath of death; *rememble*, a false memory; *macarism*, taking pleasure in someone else's joy; *musiphobist*, a person with a deep and sustained fear of poetry. Sixty-two words in total.

Next, I cut up small cedar blocks, stenciled a letter on each one, and scurried around New York City, assembling the blocks to spell out my rare words and taking photos of them—on bridges, trash cans, mossy rocks in Central Park. Last, I printed out each photo, glued it onto one of the three-by-three-inch reclaimed wooden squares, stuck a magnet on the back, and—presto—"reclaimed words on reclaimed wood" magnets. Every word came with a definition and a history of the wood.

These trinkets were quick to make and easy to sell. At five bucks apiece, we sold thousands each year and raked in cash. We worked the Christmas market in Union Square and sold in Manhattan parks, setting up at 3:00 a.m. to carve out a few feet of sidewalk, often getting chased off by the police. But it was worth it. Five thousand magnets a year at five dollars apiece—that adds up. The local section of *The New York Times* ran a story, as did the *New York Post* and *The Village Voice*. They called it word art; I called it the best hustle in town. Within a few years, we were making more than a hundred thousand dollars a year on the streets.

Through all this, things weren't going well with my girl-friend, and I began coming home less and less, spending more and more time in Brooklyn. I had found a cheap apartment in Clinton Hill—a brownstone, beautiful in its day—that had once been a frat house for Pratt's School of Design students; I rented it for four hundred dollars a month. The brothers had let it fall into disrepair. I was the only tenant; the rest of the four-story house was abandoned, caked in filth, a refuge for insects and rats, pipes leaking stains down the walls. But for me it was a five-star hotel. I painted my room industrial orange, imitating the brightly colored houses back home in Newfoundland. The romance was lost on others. One time, I brought back a pixie of a woman from a bar, and she turned to me as she walked up the treacherous stairs and asked, "You don't pay for this place, do you?"

The wood art racket paid the bills, but most of all it taught me the art of selling, which, little did I know, was going to be the catalyst of my future success in the food sector. I learned that people buy stories, not stuff. I learned to talk nonstop, reciting my reclaimed words like poetry. People would gather around my street table just to listen. Yes, they got a kick out of the knickknacks, but what they took home—what I was selling—was a story. The magnet itself mattered little; what sold was packaging a tale that they could take home and tell others. Mine was about the marriage of words and wood. It sounds cynical, the story being more important than the product, but I don't think so. It shows that we are narrative creatures, driven as much by heart as by head. I like that.

CHAPTER 10

Ocean Rescue

In the early 2000s, word leaked that Branford, the next town over, was opening up new oyster grounds in the Thimble Islands for the first time in 150 years. It was a plan hatched by the local shellfish commission, a gaggle of old men who still remembered the storied days of the Thimble Islands. Those old fellas saved my life, opening up a wormhole back to the water.

The Thimbles are a famed cluster of islands in Stony Creek, a coastal village just down the road from my trailer and a fifteen-minute drive up the shoreline from New Haven, Connecticut. They are named after the thimbleberry, a black raspberry that once grew wild but has disappeared from the shores.

A receding glacier left behind this spill of islands: massive granite knobs, stepping-stone slabs, and submarine boulders and ledges, some of which only appear at low tide. There is Money Island, Little Pumpkin Island, Cut in Two Island,

Mother-in-Law Island, Hen Island, and East Stooping Bush Island, among many others—between 100 and 365 (depending on the height of the tide, how you define an island, and if you cherish the idea of an island for each day of the year), twenty-three of them inhabited by people during the summer.

The rock of these islands is valuable. For more than a hundred years, the Totoket Granite Company dug here for the pink-hued granite that was considered the highest quality in the United States and was used for the foundations of the Statue of Liberty and the U.S. Capitol Building. The Stony Creek Quarry continues to ship stone to this day.

The islands' history tracks cycles of lawlessness, work, weather, and wealth. Captain Kidd set up on what was later named Kidd's Island and launched raids against New York–bound ships. It's rumored that Kidd hid some of his treasure on Money Island. A few Spanish gold coins have been found, enough to attract an honest-to-God treasure-hunting ship in the mid-1990s.

By the 1890s, the Thimbles were branded the "Newport of Connecticut," with President Taft summering on Davis Island, and wealthy families building swimming pools, a golf course, and tennis courts on their private islands. Artists followed the wealth. Orson Welles showed early films at the puppet house. Ayn Rand and her husband set up shop in the Thimbles and used the quarries as the backdrop for her tooth-and-claw novel, *The Fountainhead*. Greta Garbo, Mark Twain, Sinclair Lewis, and Jack London also summered at local hotels. General Tom Thumb, a dwarf who achieved fame as a circus act with Barnum & Bailey, courted a local woman who lived on Cut in Two Island.

Watermen navigated around yachts, landing hundreds of thousands of flounder, stripers, eels, and blackfish. The protected waters, salt marshes, and granite outcroppings made fertile grounds. Menhaden were caught for fertilizer. Each season brought different bounties: flounder, mackerel, blackfish showed up in April; by July, snappers and porgies were in season; winter was for eels.

But shellfish were the Thimble Islands' true claim to fame. Due to the tasty mix of swirling fresh and salt waters, and strong inland tides and currents, the clams were sweet and the oysters sharp and briny. The Mattabesic tribe were the Thimbles' original shellfishermen. Archaeologists have found shell piles that were left three centuries before settlers showed up. The Mattabesic slurped them raw, roasted them in fire pits, and steamed them in seaweed wraps.

Shellfish were also the lifeline for salt marshes. Trillions of oysters were once the cold-water coral reefs of America, functioning as the foundation of the ecosystem. Beyond offering refuge to hundreds of marine species, one acre of oyster reef filters 140 million gallons of water an hour and removes three thousand pounds of nitrogen a year. When the first English settlers landed, oysters were filtering the entire bay about once every four days. Ships had to navigate around the massive oyster reefs.

And our favorite mollusk spawned the first wave of green jobs. By 1880, there were 52,805 workers employed as harvesters, dockworkers, shuckers, and canners. The first black business owners were oystermen on Staten Island. Harvesting began with locals wading into estuaries to hand-harvest, then tonging from sail-powered skiffs, and eventually it grew into large-scale dredging operations. From 1880 to 1910, Americans pulled in more oysters than all other countries combined. Shucking contests became a competitive sport and drew huge crowds; one person opened a record twenty-three hundred oysters in two hours, eighteen minutes, and nineteen seconds—which is insanely fast. Even the best shuckers can't open half that many in two hours.

It won't surprise you that "progress" ruined nature's bounty. As early as 1715, government officials banned harvesting during the months whose names contained "r"s, because they were worried about overharvesting. But the game changer was the invention of the dredge in the late nineteenth century. Commercial dredgers scoured the bay bottom, and soon "oyster wars" broke out. Fleets from New England charged down the coast, pillaging and poaching along the way. Oystermen began murdering to protect lucrative harvesting grounds, which in turned spawned new police departments dedicated to keeping the peace. Ignorance of food safety risks sparked cholera scares, which temporally ruined the market and triggered a massive drop in sales.

But it was pollution that brokered the truce. By the 1920s, as cities became more crowded, sewage poured out of urban rivers, and diseases spread from bay to bay, killing off billions of oysters in short order. East Coast oyster populations plunged to less than 1 percent of historic numbers. It's impressive how quickly humans can wipe out a resource-dependent economy. By 1930, *The New York Times* was reporting, "Oysters, once plentiful and considered a frugal repast, are gradually being classed as luxuries and will soon become a delicacy."

With the passage of the 1972 Clean Water Act—arguably the most effective piece of environmental legislation in U.S. history, and passed during the Nixon administration—oysters returned to some of the embayments along the East Coast, especially on my home water, Long Island Sound. Norm Bloom & Son, who work more than fifty thousand acres of Connecticut waters, are largely responsible for the revival. For years in the 1980s, they tirelessly seeded beds, until Long Island Sound was again one of the most productive oyster areas in the country—albeit at a fraction of its nineteenth-century levels.

LIVING IN THE HALF-SHELL

With the new Thimble Island leases available, I decided to hitch a ride on this history and remake myself into an oysterman. At the time, I viewed it as second fiddle to commercial fishing: an inferior but necessary choice, because I was trapped in the suburbs; my last-ditch effort to stay on the water. Instead, it became my life's work.

First, I had to get myself a piece of water. I headed to the shellfish commission meeting held at five o'clock on the second Tuesday of every month. The second I walked in, I was at home: the stink of mud caked to rubber boots, arms crossed in antigovernment defiance, the tensions of fratricidal feuds over poaching and cut gear. The fingers of the suburbs released their chokehold, and I could breathe again.

The first order of commission business was a prison letter from a guy requesting to lease grounds. It was read aloud, and his lease was approved.

Next up was a David-versus-Goliath battle between local shellfishermen and a multimillion-dollar energy company that wanted to run gas lines through the middle of shellfish grounds. The company offered a fifty-thousand-dollar buyout. No takers; company lost. Sociologists call it a "sticky" profession; it's water, not money, that keeps watermen working. My lease came up for a vote, breezed through without dissent. It was twenty acres and cost fifty dollars per acre per year. A thousand bucks a year for twenty acres—try to get that on land.

This whole idea of "owning" water was alien to me. On the

Bering Sea, we fished wherever the fish were—even illegally in Russian waters. As lobstermen, we set our traps at will, limited only by turf wars. Now I was paying for a little unmarked patch of Long Island Sound. Already, I was feeling claustrophobic, like I was getting boxed in.

But as I soon learned, "owning" the water didn't actually mean I owned the water. I wasn't actually going to own my farm; I only owned the right to grow shellfish or other permitted species. This was all hard to get my head around. Unlike on land-based farms, there was no concept of trespassing, so the community was going to be able to do whatever they wanted on my "property." Anyone was going to be allowed to boat, fish, swim on my grounds. The only thing the public was blocked from doing was farming. Commercial fishermen could even still gillnet, and lobstermen could still drop traps. As a farmer, I owned a process right, not a property right. This ocean-farming world was going to take some mental adjustment.

I bought a boat, named it *Mookie,* after both Mookie Wilson of the Mets and my black Lab. It was a blood-red twenty-two-foot Midland I picked up for twelve hundred dollars in Wickford, Rhode Island. Total piece of shit: narrow beam, engine billowing toxic white clouds whenever I fired her up, the fiberglass floor soft from wear and rot. She could only handle two people before the scuppers began to sink below the waterline.

I showed up on the docks, kept my head down, and earned the nickname Mookie, after my boat. I'd spent enough time in small communities to know that I needed to come in slow and quiet. One of my first days on the docks, I ran into a brick shithouse of a man. I'd seen him running an aluminum boat around the Thimbles, moving barges, dumpsters, cranes on and off the islands. The man had mad skills, could thread a needle when it was blowin' twenty knots.

"Where you from?" he shouted at me.

"I'm living over in Guilford," I said.

"No, where you from?"

"Well, I was born in Newfoundland."

"Yeah, me too." You gotta be kidding me. I was working the same docks with a Newf.

"What are the chances that we'd both be in the States working here!"

"Chances are good," he responded. "We're everywhere!"

Typical Newfoundlander. Worldwide, we find each other.

Carl's story was common. He was one of seven children, grew up in a small outport, and dropped out of high school to go fishing. After the fish stocks crashed, he came to the States to look for work. He ended up at a construction company that built and remodeled homes on the Thimble Islands. As a fisherman, he didn't have construction skills, so he ran the company boats, pushing their hulking steel barges.

Drive a Newf off the island and he'll walk the globe lookin' for water.

Carl and I would always hit the same notes when we saw each other on the docks.

"Whattya at, b'y?" he'd ask. Our typical Newfoundland greeting.

"Nothin', b'y, what ya at?"

"I'm heading back home soon. Going to get my yearly moose and see the family."

Carl and others off the docks were good to me—helping where they could, and always peppering me with questions about what in the hell I was trying to do underwater. Took me quite a while to figure that out.

I named my farm the Thimble Island Oyster Company. With lease in hand, I did what any novice fisherman or farmer should do: shut up and listen. It was the early 2000s, and there were local watermen who had been in the trenches for twenty years innovating new ways to grow oysters. There was Robert Rheault at the East Coast Shellfish Growers Association, Tessa Getchis at the Connecticut Sea Grant, Karen Rivara at Aeros Cultured Oyster Company, and the how-to manuals of Dale Leavitt at Roger Williams University in Rhode Island. No way I could have done it without these folks.

It turned out there were a dizzying number of ways to farm oysters. Some used rack-and-bag culture, with rigid grow-out bags tied to steel rebar racks. Others floated their bags on the surface or used stacked trays. But I elected to go with cages. My lease was in deeper waters, so cages were a good fit. Each cage had shelves with mesh bag inserts, with a rope bridle and a line up to the surface attached to a buoy. Plus, hauling cages was similar to hauling lobster traps, so more in my fisherman's comfort zone.

I picked up my first batch of forty thousand seed from the New London Ferry, packed in burlap bags and shipped over by Steve Malinowski from the Fishers Island Oyster Farm—a long-time hero of the industry. It was late in the day, so my plan was to keep them in my truck for the night, then bag and set the cages the next morning. But around 6:00 p.m. the weather turned, and air temps began to plummet toward freezing. I made a frantic call to the hatchery and confirmed that if the seed froze they'd die. I was still living in the Airstream trailer, so I dragged a kiddie pool inside and piled the seed up high. My propane heater had been broken for months, so I set up an electric heater next to the tub of seed, powered by an extension cord. I draped wet towels over my seedlings.

My girlfriend was sleeping as I sat next to the pile, on the trailer floor, nodding off, waking in starts, turning the heater on and off, wetting towels. She woke, groggy and grumpy, growling, "What the hell are you doing?" I had no fucking idea what I was doing.

In the morning, I hauled the seed to the docks, cut the top off a gallon bottle of Coke, counted out a thousand seed per container, and poured them into my rigid mesh bags. I loaded fifty bags into three-racked cages, and dragged them onto *Mookie.* Her guts were so waterlogged that when loaded down with cages she had the stability of a canoe. I started up my outboard; white smoke billowed and fogged the docks. A drunk rushed over and put on a show, wildly playing air guitar in the toxic fog machine.

Again, my timing was bad—the wind and waves picked up. It was just a ten-minute ride, but my little boat got beat up bad on the way out. I'd worked big boats my whole life, but this leaky tub made me wish for a few minutes that I knew how to swim. I rushed to set my cages, dumping them all off the gunnel. Not one set correctly.

That first year, I thought I was growing oysters, but in fact I was running a death camp. Forty thousand died in the mud. The cages got too heavy and tipped over, and everyone drowned. I was devastated. What a new feeling. I'd never felt attached to the fish I'd caught on the trawlers—the ocean grew the fish, I didn't. But these critters I had bought as seeds and raised. They were my brood, and I had committed filicide. The few that did make it through I brought to a local chef for a taste. When he tried one, he immediately spit out the muddy meat. "This is disgusting," he told me.

I was a shitty farmer.

GROWING A BLUE THUMB

I soon learned that ocean farming had zero relation to fishing. As fishermen, we are the last hunters—we search, chase, kill. It's about highs and lows. It makes us hard-driving, hard-living humans. Now there was no hunt, no high seas, no wildness. The shift to the flats was a blow to the saltier side of my ego.

Except for lashing up to the dock and handling a swell or two, my high-seas skills and temperament were of little use. Instead of living for the thrill of the chase, I was now an underwater farmer, living a small, modest life. Everything was different—the routine, the weather, the physicality, the risks and rewards.

I was moored to my tiny twenty-acre ocean plot, no longer free to roam the eight hundred thousand acres of the Bering Sea. No rogue waves—a six-footer is the thrill of the year—or strange monsters pulled from the depths. Just a gentle little ten-minute boat ride to the oyster grounds, followed by a day of hauling cages. Put the buoy line in the hauler, crank up the cage, set it on the gunnel, open it up, take out the mesh bags, dump the oysters on the table, sort out the dead ones, put them back in the bag; bag goes back in the cage, cage goes back in the water. Do it again, and again and again.

My nervous system began recalibrating; I found new wells of pleasure and meaning in the meditative, repetitive, and solitary. I loved the winter months: yachts hauled out of the water, rich kids gone back to private school, cold keeping tourists off the docks. I loved breaking ice with a sledgehammer to get out of the harbor, and days when I was blinded by snow. I loved the early-morning rifle crack of duck hunters, and the winter seals on the rocks keeping watch over my farm.

The work also gave me a community at a moment when things weren't going so well in my personal life. My girlfriend and I were on and off, until we were off for good. I moved into a beat-up old cabin near the farm, and spent as much time as I could at the docks. Watermen don't "hang out" per se, but we spend an awful lot of time milling around the docks, passing beer and stories.

And I did get to go to battle sometimes. I had buoys cut as a shot across the bow for the new kid in town. One whacked-out waterman reported me to the FBI and the Department of Energy and Environmental Protection (DEEP) for supposed poaching. Another accused me of colluding with Google to block his website from searches. Although pretty mild compared with my days in Lynn, the skirmishes at least spiced up my days. I'll take what I can get.

My mom hated my winters. She cursed my refusal to learn to swim and knew that at times I had to hide in the dark of the bow to quiet my epilepsy. Finally, she called. "I can't take it anymore." I knew what she was going to say. "I can't take you out in the winter, alone on the boat. All that ice and snow, and you can't even swim. I've listened to the marine forecasts and dreamed of you lost at sea too many times. I'm too old for this."

"What do you want me to do about it?" I asked.

"Hire help," she said.

She had a point. My boat was a broken-down, waterlogged, barely floating hulk. Many days, *Mookie* couldn't even make the round trip to the farm. It got so bad that I'd tow my thirteen-foot skiff out with me in the mornings; when *Mookie*'s engine choked and stalled, I'd haul her home with the skiff. One time when I was pulling oyster gear on *Mookie,* I had five or six hundred pounds of traps stacked on the deck. I looked down; water was ankle-deep and rising in the stern. I shoved the cages back into the water and threw *Mookie* into gear. The forward momentum began to drain water out the rear scuppers. It was November, and water temps were in the forties, so I decided it was best to head back to the docks. The water level continued to drop, but then my outboard stalled. Water came pouring back in. I was sinking fast. No other boats were around, and I didn't have time to radio the Coast Guard. The engine started back up; I threw her in gear and made it fifty yards more before she stalled out and began sinking again. I radioed to Gerard, who ran Thimble Marine Service on the docks. "Engine keeps stalling, I'm taking water fast. I'm going to try to get to the boat ramp, so have a trailer ready."

"Roger," said Gerard, always game to save my ass. Drive, dump water, stall. Drive, dump, stall. Happened four more times, but I got *Mookie* to the ramp, and Gerard pulled her out before she sank for good. I was just as shitty a captain as I was a farmer.

The first guy I hired was a hulking ironworker. He must have topped out at 275. Pure muscle and pride. I hired him for twenty dollars an hour, which was crap wages for someone who'd been working high steel for union wages. But he wasn't out there for the money—it filled the time between construction gigs and got him out on the water every day. He was a tough fucker. My hauler had an electrical short somewhere, and he'd sometimes get shocks run up his arm while hauling cages out of the water. Didn't faze him; he thought it was funny. When I hired him, we didn't negotiate over wages, hours, or benefits. He wanted to drink while hauling cages and began by making his opening bid to start drinking at 6:00 a.m. We settled on his opening his first beer at 9:00 a.m., which I saw as a more reasonable starting time. One day, after a few months of working together, I was gobbling down a breakfast sandwich between cages when a piece of roll got caught in my throat. I sputtered, choked, and grabbed his bottle of water. Took a gulp and choked even worse. It was pure vodka. Turns out the negotiations were for when he was allowed to start drinking beer; vodka began before we left the docks. I gave in, and most days I joined him.

After my first failed season, I tried again and worked hard and vigilantly. Each week, in rain, shine, or blizzard, I hauled my seed onto the deck to sort out little from large, grading them on quality and shape. The critters did well, and after about five months, they had reached one and a half inches. To encourage tougher shells and deep cups, I'd toss them, five hundred at a time, into a rotating metal cage I built, designed to break off the edges and toughen the shells. I scraped the slipper shells off the oysters and seaweed off the cages with metal brushes, then sent them back to the bottom. I did this routine year-round, hauling hundreds of cages until the oysters were fully grown and ready to harvest.

Slowly but steadily, I developed a blue-green thumb. My seed grew quickly, beautifully infused with the merroir of the Thimble Islands. They had deep, golden shells, and the meat was plump; the liquor, clear and sharp, hit the palate with the full force of brine, finished by a minor chord of sweetness. My first check was from Bud's Fish Market, located a few miles down the road. It was only $150, but I have a picture of myself holding that check with a beaming smile, the newly discovered pride of selling something I'd grown. I liked the feeling.

My farm attracted mysterious and tasty delights. Little translucent pea crabs hid in the gills of my oysters. They were crunchy, a sweet-and-salty mix of flavors. I dug up a 1913 *New York Times* article entitled "Oyster Crabs—The Epicure's Delight," which explained that these "highly prized and correspondingly expensive" crabs are so tiny "that it doesn't seem as if you were getting much for your money. . . . What you do get is a sublimated reminder of the daintiest shrimp you ever ate, with about the same relationship in flavor that a mushroom has to a Brussels sprout."

There were toadfish, or "oyster crunchers," which grunted to attract mates and were sent into space by NASA for microgravity testing. Often I'd fillet the two secret slices of white meat for lunch. I learned about the strange lives of the slipper shells, which are sequential hermaphrodites that attach to oyster shells and stack high, up to twelve on top of each other—with females always on the bottom of the pile and males always on top. When a new male begins growing on top, the male underneath turns into a female. Yes, they were perverse critters, but their fertile meats tasted like sweet little clams.

It was a dirty job. I'd return to the docks covered in stinking black mud. Tiny skeleton shrimp wiggled in my beard. One time, when I was stacking gear in the back of my truck, getting ready to head home, a crowd of elementary school kids piled off a bus and streamed toward the tour boat. One of the little girls came over to me.

"What are you doing?" she asked.

"Oystering," I said, and pulled a few from the cooler to show her.

"Do they bite?" Cutest question ever.

"Nope," and I gave her one to hold. A couple other kids split from the group and came to look. She had it in her hand, holding it like it was fragile. She looked at me with disgust, wrinkled her nose, and said, "You smell bad!" One of the other kids leaned in for a sniff and said, "Ewwwww! He smells!" Then the other kids chimed in, yelling "Ewwww!" I shrugged, saying, "This is what work smells like." They scurried off and joined the rest of the class.

Oyster farming also turned out to be everything industrial aquaculture was not: sustainable. I learned that oysters are powerful agents of restoration: they filter up to fifty gallons of water a day, pulling nitrogen out of the water column. In excess, nitrogen triggers the oxygen-free dead zones that are spreading throughout the globe. Recent work done by Roger Newell of the University of Maryland shows that a healthy oyster habitat can reduce total added nitrogen by up to 20 percent. A three-acre oyster farm filters out a nitrogen load equivalent to what is produced by thirty-five coastal inhabitants.

Soon I found myself embedded in a new community. I started the first shellfish community-supported fisheries (CSF) program in the country. Just as in a land-based community-supported agriculture (CSA) program, the idea was that people would buy a share of the farm at the beginning of the season, which would provide the up-front capital to allow us to grow food for them. In exchange, every month, they'd have a box of shellfish waiting for them on the docks. It was popular, and shares would sell out each season in a couple hours.

People loved our oysters, and a community grew around the CSF. Members would hang out for hours on CSF pickup days sharing recipes, or bring along relatives visiting for the weekend. When word got out, people started flooding to the docks, wanting farm tours. And they brought me food of their own: eggs, pickles, spiced vodkas. It seemed everyone was running experimental kitchens at home and wanted to share. I posted a steady stream of photos showing members plating their shellfish meals. Fishing 2.0.

I had been welcomed by the new urban class of "foodies," a strange, ritualized culture marked by the trancelike state they'd go into after the first bite of a new dish. A slight smile curled onto their faces as the oyster liquor hit their taste buds. Their eyes would close. A moment of silence. Then a practiced attempt at poetry, as they detailed the swirls of flavor. Never one to fetishize food—I still ate at the gas station most nights— I found this new and, at first, alienating. But, God, how they loved my oysters, my pea crabs, my slipper shells. I'd quickly become proud of the food I grew, and adopted their culinary dialect.

Although the skinny-jeans hipster crowd are easy targets, for a farmer it was fertile ground—even more so for an ocean farmer. We were celebrated for the dirt under our nails, or, in my case, the dried mud and salt in my beard. This was the big tent generation, overflowing with solidarity for everyone up and down the food chain. Foodies showed up on picket lines for striking fast-food workers and opposed the deportation of undocumented farmworkers. They clamored to meet their fishermen, farmers, and shellfishermen. I may not have been a foodie myself, but these environmentalists, artists, food desert activists all welcomed me with a warm embrace, inviting me to meal after meal, authentically, generously interested in my work. Off the docks, I had found a second family.

Building Your Ocean Farm

For more than a decade, I've been working to drive down the cost of ocean farming. We have two advantages over land-based farming. First, seaweed and shellfish just eat what's in the water, so we don't have to feed, fertilize, or water our crops. That keeps fixed overhead costs low. Second, we don't have to deal with much gravity underwater, which means all we need to use are ropes and buoys, and some anchors. This keeps start-up costs wickedly low.

So if you can scrape, borrow, or steal twenty thousand dollars and an old boat, you can start a farm. I did a Kickstarter campaign and sold knickknacks on the streets of New York so I didn't have to go begging to the banks.

The farm itself is simple: just anchors, ropes, and buoys creating a grid system. The anchors sit on the ocean floor along the edges of the farm, spaced twenty-five feet apart. Anchor lines are attached to each anchor and run vertically up to the surface, where they are attached to an eighteen-inch buoy, also called a float. Then, six feet below the surface, horizontal ropes are strung between the anchor lines. The horizontal lines are where you will attach your crops, which grow vertically down into the water column—seaweeds, scallops, and mussels. Every fifty feet, twelve-inch floats are attached to the horizontal lines. That's it! This is the framework for the farm. All you'll see from the surface is a gridwork of bobbing buoys, which is exactly what we want. Ocean farms are visually low-impact and look similar to a boat-mooring field.

When I was in Wando, South Korea, for a meeting of seaweed farmers, there was an ocean-farming hardware store every five blocks or so. Not a hardware store with a section dedicated to ocean farming—I mean a whole store! Glad the airlines don't let you take buoys as carry-ons; otherwise, I would have blown a few paychecks. We don't have corner stores for ocean farmers in America, but you can find and make what you need.

On the water, rope is called "line" so call in an order for half- and three-quarter-inch sinking anchor lines. (There are also floating lines, so make sure to specify "sinking lines.") The three-quarter-inch lines are the vertical anchor lines that run from the ocean floor to the surface, so the number of feet you'll need to order will be based on the water depths of your site at high tide. (You can consult paper or digital charts published by NOAA, to find the depths at your site.) The half-inch lines are the horizontal ropes that run below the water. These are five hundred feet long each, so you'll need enough for however many lines you plan to grow. Next, order floats—enough for ten per horizontal line. These could be lobster buoys or aquaculture floats—anything that will hold the farm afloat will do.

The last items to buy are navigation buoys. The safe boating commission in your state will dictate the size of these, but I use five-foot-tall buoys, twelve inches in diameter. They need to say "Danger: Gear Area" with a reflective diamond, which is the international marine sign to alert boaters to beware. Although the navigation buoys are required by law for anyone farming the sea, your farm is not going to block most inshore boaters, because the rope scaffolding begins six feet below the surface. Most inshore boats draw less water than that.

Now it's time to build your anchor systems, which will hold the farm in place, even during big weather. Each horizontal line gets two anchors on each end. If you've got the funds, buy mushroom, pyramid, or screw anchors. I make my own by building a concrete form out of three-quarter-inch plywood in the shape of an upside-down pyramid. Make four of these, which will allow you to create a full anchor system at once. Pour four eighty-pound bags of concrete mix (about $4.50 a bag) into each form, and sink an old piece of rebar or a couple links of chain into the wet 'crete; this is what you'll tie your anchor lines to. Let them set for twenty-four hours. It's key to make sure the anchors are light enough so that two people can load and set them from a small skiff and you won't have to hire a mooring company, which can cost up to fifteen hundred dollars a day.

Time to rig and set the farm. You can pull this off in one day if you've planned well. Cut your three-quarter-inch anchor lines to the right length (based on the depth you figured out earlier), adding a few feet to allow for knots. Take one end of the anchor line, pass it through the rebar or chain link on two of your anchors, and tie it off with a common anchor hitch, which is designed to keep chafing at a minimum. If you used chain link, you could also shackle the chain together and tie the line to the shackle with an anchor knot. What you want to end up with is two of the anchors roped together as a unit. Tie your eighteen-inch buoy to the other end of the anchor line with a bowline knot. Now you've got one anchor system. You'll need two of these per line, one for each end.

Next, head out to the farm and set the anchors. This basically involves carefully sliding them overboard off the gunnels of the boat, while being extremely careful the lines aren't wrapped around your hands or feet—otherwise, you'll become fish bait. Set one of your anchor systems every twenty-five feet. Then head to the other end of the farm and set a second line of anchors. They should be five hundred feet apart. Finally, string the five-hundred-foot horizontal lines between all of the anchor systems—these are the "longlines." I use a figure-eight knot in the anchor line and attach the horizontal line with a bowline knot.

Now it's time to purchase the prepping gear for shellfish. For mussels, you'll need "socks" that come in large rolls of mesh tubing that you stuff your mussel seed into and tie off to the longlines after kelp season. For scallops and oysters, you need lantern nets, which are ten-tiered mesh columns that hold shellfish and hang vertically down from your horizontal longlines. I also use oyster cages, which are basically just modified lobster traps with shelves to hold mesh bags of seed. I make them myself, but for your first year, I suggest using lantern nets, to keep costs down and make farm maintenance simple.

That's all there is to it. Farm's built, and you can start calling yourself an ocean farmer.

part **3**

CHAPTER 11

Up from the Depths

Then the storms hit.

I had been farming oysters for almost a decade, and was beaming and proud of having created a handful of jobs and filled folks' bellies. But on August 28, 2011, this all came to a screeching halt. Tropical Storm Irene marched up the Eastern Seaboard and destroyed my farm. She attacked with a six-foot storm surge that barreled across Long Island Sound and buried my oyster cages in three feet of mud.

Over half of my gear was lost at sea. Eighty percent of my crop destroyed. Hours after she blew through, I jumped on my boat to survey my grounds. The seas were still rough, too rough, as I hauled my cages, which were swamped with more than a thousand pounds of mud. Each cage I hauled: thousands of oysters dead. The tailwinds of the hurricane circled back around, but I kept hauling, looking for some signs of life. My steel davit snapped under the weight; then my engine went cold. The swells were now in full force—Irene lingering out of pure meanness. I threw the anchor, but she didn't catch, and the weather blew me hard onto the rocks. I called out an emergency May Day, and the Coast Guard hauled me to safety after a couple hours of chill and fear. I had done a stupid and desperate thing to try to save my beloved farm.

Irene was ranked the seventh-costliest storm in U.S. history, and I was counted in the damage tally. My losses added up to over a hundred thousand dollars. No insurance, no government aid. I had no option but to take the body blow, stop whining, and rebuild. Like farmers up and down the East Coast, I began again, dumping thousands of dollars into buying new gear, new seed. Starting from scratch.

Then it happened again. Only a little over a year after Irene, another storm, Superstorm Sandy, rained hellfire on my twenty acre farm. She left not a single oyster alive and destroyed even more gear. This was the end of the line. It had taken me years to build out my patch of ocean, but just a handful of hours for nature to destroy it. After Irene, I had put my head down and accepted my fate. It was heartbreaking to lose everything, but hurricanes of Irene's force were rare, so I figured I could take the hit and rebuild. Now, this second time around, I felt helpless against the increasing wrath of extreme weather. I had remade myself from a pillager of the high seas into a blue-green farmer, but my farm, my oysters, and my livelihood had become a canary in the coal mine for a climate crisis that arrived a hundred years ahead of schedule.

The day Sandy hit, all the mind-numbing politics and shrill debates as to whether climate change was real came to an end for me. Two farm-destroying storms in two years was evidence enough that I was living in the new age of climate crisis. News began pouring in about other escalating threats to oysters: acidification both undermined the ability of shellfish to produce calcium for their shells and was already killing off millions of seed on the West Coast. Disease outbreaks of dermo and *E. coli* were flaring up more often as water temperatures broke records daily.

And it wasn't just my farm. I began talking to marine biologists and climate activist groups like 350.org and learned that the unholy trinity of climate change, acidification, and overfishing had our oceans locked in a death spiral. Lobsters literally vanished overnight in Long Island Sound. Globally, the average body weights of six hundred species are expected to shrink 20 to 30 percent by 2050. Acidification is climbing ten times faster than it did during the mass marine extinction of fifty-five million years ago. Dune Lankard, an Alaskan fisherman and founder of the Eyak Preservation Council, told me that 2018 was the worst salmon season for Native Alaskans in a hundred years. The Gulf of Maine is warming faster than 99 percent of the world's oceans. All this is taking place at a moment when 90 percent of the world's fish stocks are fished at unsustainable levels, and nitrogen pollution is spreading hundreds of oxygen-free dead zones around the globe.

Here's what the Big Greens like Greenpeace don't get: For most working Americans, climate change is not an environmental issue. It's not about saving birds, bees, or even bears stranded and starving on melting ice floes. It's about bread-and-butter economics. In the wake of Sandy, eighty-three thousand people lost their jobs in New York City. Unemployment rates skyrocketed along the storm path up into Vermont. In the last ten years alone, the impact of climate change has cost the American economy at least $240 billion a year, and the future economic costs within the U.S. borders are predicted to be the second-highest in the world, behind only India.

This is not a crisis to be faced by future generations. It's not a slow lobster boil. According to a 2018 United Nations report prepared by ninety-one scientists from forty countries, by the year 2040 we should expect a world of swamped coastlines, rising poverty, worsening food shortage and wildfires, and mass die-offs of coral reefs. Yes, that's twenty years from now. The findings surprised even the authors of the report, with one telling *The New York Times* that it was "quite a shock . . . We were not aware of this just a few years ago."

The more I learned, the more I realized my farm was just the tip of the iceberg for a climate crisis that is gutting the American economy. Greenpeace can worry about the polar bears; I'm worried about how we're all going to afford to put food on the table.

THE REBIRTH OF THE OCEAN FARM

The minutes and days after Sandy were emotional and heady. I had no hint that this was the great turning point for me. I felt cold and lost. And alone. Ocean farming had rooted me, gifted me with a new identity. With the farm gone, so went my sense of meaning. I was poor and adrift. Fucked in the suburbs once again.

Ironically, I even became allergic to my oysters. I was using my table saw one day to cut a new bunk rail for my skiff trailer. The saw kicked; the plank was launched at my head and hit me in the nose and forehead. I was in and out of consciousness, blood everywhere and chunks of wood buried deep in my face (some of which remain today). I have memories of the doctor yanking wood out of my face with some kind of pliers, pulling so hard my head was lifting off the table.

At the hospital, they told me I had two cracks in my skull and a bubble was forming in my brain. I don't really know all the procedures they did, but they pumped me full of some kind of dye to do a test, and, next thing I knew, I went into a full-on allergic reaction. A few weeks later, when I was finally released, I went back out on the boat and cracked open a few shellfish for breakfast. My throat closed and eyes swelled. I was fucking allergic to my own crop now.

Hope doesn't require cheery roots—it prefers the dark soil of failure to sprout. I had taken many body blows in my life: as an out-of-work fisherman, as a failing student, now as a struggling oysterman. When I look back at those heavy, despairing days after Hurricane Sandy, they strike me as some of the most creative and thrilling days of my life. I'll go further: Sandy is the best thing that ever happened to me.

The disaster forced me to reimagine the way I grew food in order to keep my business alive. Climate change was real, and if my farm was to survive, I needed to become a climate farmer, growing climate cuisine—adapting and learning from the changing seas. Climate crisis demanded the cultivation of native species and new farm designs that were more resilient to rising water temperatures, acidification, and extreme weather. It demanded a new locally grown ocean cuisine that was delicious and nutritious, but still flew the oyster's banner of sustainability. My whole life, I had been a poor student, chafing at classroom settings, but after Sandy, I was to hit my stride and begin my real education. I knew this new phase was going to take some serious bootstrapping, but I was a worker, determined to make it work.

So I hacked my farm to survive the coming storms.

First I had to do a full redesign. I humped back to the Airstream and got to work. My sketches and thinking moved in fits and starts; often decisions had unintended effects in practice. I borrowed every idea I could from a thousand years of ocean farming. My goal was to synthesize the good, refuse the bad.

Clearly, it was time to go vertical. In the wake of Sandy,

my farm had to evolve to avoid the storm surges that rolled through during hurricane season. I also needed the flexibility to raise and lower my gear to adapt to ever more volatile weather conditions. I started sketching out rope-scaffolding systems, modeled partially on mussel farms in Spain, to pull the oysters off the floor. To do this, I needed anchors that would withstand heavy winds. My farm was exposed to southwest winds, and the long fetch brought through rollers that would drag my gear into the rocks. So, for anchors, I created a mold for upside-down pyramids that cost only thirty dollars apiece to build—a couple of bags of concrete and some chain. A vertical line was tied to the anchors and ran up the water column to a surface buoy, which held a relatively low buoyancy of sixty pounds. This minimal buoyancy was vital for mitigating the impact of hurricanes. Keeping the buoyancy low allows the buoys to give way and submerge when a storm rolls in. Be a willow, not an oak: don't try to fight the sea; bow and step aside and let her rage and roll through.

I sank two of these anchors five hundred feet apart. Next, I strung a horizontal line between the two anchor systems, using sinking line to hold it six to eight feet below the surface. To keep everything in place, I attached smaller buoys to the horizontal line every fifty feet. Put twenty of these longlines in a grid and you've got yourself a good-sized farm.

I needed to ensure that all the surface lines would stay tight even when the tide dropped—which in the Thimbles is six feet. Otherwise, the horizontal lines would begin hitting each other and the buoys would tangle. So, halfway down each anchor line, I attached a twelve-inch buoy to keep the vertical lines tight regardless of tidal changes.

One of the benefits of the rope-scaffolding design was that I could sink the farm deeper into the water column during storms to avoid heavy seas. This setup also allowed me to find the nutrient sweet spots that would stimulate growth at different levels. I could loosen the lines, remove most of the floats, and drop the farm far below the surface, thereby mitigating storm damage and growing a wider variety of species. There were other benefits as well. This system was simple, fast, and cheap. All I was using was line, buoys, and concrete—no pulleys, pens, feeding machines, cranes. The whole farm could be installed in three days, and my back-of-the-envelope calculation of costs was between ten and twenty thousand dollars. Good luck trying to build a land-based or salmon farm that cheap.

It was also low-impact. Rather than building massive fish pens, vertical farming is designed to tread lightly on the ocean commons. Because it was mostly underwater, from shore there would be minimal aesthetic impact. And the farm carried only a twenty-acre footprint while producing a huge amount of food. Later, after it was built, I'd see locals boating, fishing, and swimming where I worked. And commercial gillnetters regularly set their nets around my farm, because fish are attracted to my crops; the goal was protecting, not privatizing, the ocean. As farmers, we need to be the park rangers of the ocean commons and build community support through inclusion and thoughtful design. Countless aquaculture projects suffer quick deaths when communities mobilize in opposition. We need farms that are welcoming to fishermen, both recreational and commercial. Kayakers, divers, and duck hunters need a place to play, explore, and hunt. Turns out design matters.

I also now had the whole water column to work with, which meant I could segment the farm into four sections and choose from a wider variety of off-bottom crops. There was one hard and fast rule: grow only zero-input species that won't swim away and don't need to be fed.

I had learned on the salmon farms how expensive—and environmentally destructive—it was to grow finfish, which locked the industry into a high-input, capital-intensive, and unsustainable agricultural system requiring monoculture, antibiotics, wild-fish feeds, and increasingly GMOs, if the farmer was to turn a profit. The underlying logic of the model assumes that our only option is to farm what people already enjoy eating. Just look at what was caught in the wild—cod, tuna, salmon—and build an industry around these existing markets. But the taste for these fish emerged from a wild fishery, which is a world apart from farming the ocean.

I needed a different starting point, based on the fundamental belief that our tastes must bend to the ocean's will. Regardless of existing market demand, what makes the most sense to grow in the ocean? The answer is pretty simple: grow crops that act like oysters. Restricting myself to shellfish (and later sea greens) instead of fish unlocked a farming model that is economically viable and ecologically sound. I could grow things that don't need to be fed, watered, or fertilized—in other words, things that are cheap to grow. I didn't need a Ph.D. in economics to understand that a business model grounded in zero-input crops is a good business to be in.

So I got to work choosing my crops. Clams were an easy choice. They thrive in mud—my lease had plenty of mud on the bottom—and were seeded directly on the seafloor. Next, I had to keep my oysters alive. As before, I packed mesh bags full of seed, put the bags in cages, and dropped them on the floor. But now, with the scaffolding, I could move my oysters off the bottom once they had grown to about two inches, into hanging lantern nets suspended from my surface ropes. There they could grow out to market size, about three inches.

Mussels had been plaguing me for years. Every spring, millions of seed floated through the water column and settled on my buoys and cages. I'd spend two weeks or more every year scraping them off and dumping them back into Long Island Sound. I'd never considered farming them—but I was evolving, now looking to flip a nuisance species into an opportunity. I didn't know much about farming mussels except from watching videos of large-scale operations on Prince Edward Island, Canada—late-night shellfisherman porn. So I called up Scott Lindell at the Woods Hole Oceanographic Institute to figure out how to do this. He and his brilliant number two, Dave Bailey, had been setting up farms all over the globe, from Morocco to Spain. Scott would turn out to be an important mentor for me.

Scott suggested I use my scaffolding system to string mussel socks. You make mussel socks the way you make sausages, stuffing seed into long mesh tubes. The larger operations have machines that stuff millions of mussels into socks, but I was running a DIY shop, so I used eight-foot lengths of PVC pipe to make the fifteen-foot socks. Then I brought the stuffed mussel socks out to the farm and hung them on the lines. Let them grow out for six months, and then harvest. Simple.

I'd also been experimenting with—meaning slaughtering—scallops for a couple years, never able to get them to grow in cages. But now I had hanging lantern nets. The Chinese had been growing scallops this way for decades, and it was the perfect fix—but with every fix came new problems. In the waters I was growing in, the shells grew fast, but the adductor muscles—those little medallions you all eat—didn't grow to marketable size. So I had large shells with little meat.

Good news is, Brooklynites like to eat baby everything: baby kale, baby spinach, "petite" oysters. So I began serving baby scallops at events, shucked and raw, just like oysters. Instead of cutting out the adductor muscle and throwing away the rest—which is common industry practice—we served the whole scallop, guts and all. Sometimes hustles turn out delicious: baby scallops were sweet, with a swell of surf at the tail end. They were easy to shuck and lay on the table, and the fabled ridged shell hit high aesthetic marks. Plus, they sold for a buck apiece—great for farmers.

The one missing piece was a strong winter crop. Oysters and clams head to market from spring through fall, mussels are ready in the spring, scallops in late summer and fall. I was hell-bent on farming every season. But what else could I grow?

DR. SEAWEED

I picked up the local paper, and there he was: Dr. Charles Yarish. The reporter nicknamed him Dr. Seaweed. He had started an experimental kelp farm off the coast of Bridgeport, Connecticut, to research the ecological benefits of sea vegetables. Right here in Connecticut, this tiny little state, was Dr. Seaweed! A lightbulb went on.

I pulled out my phone and called Dr. Seaweed; he answered. There's no fate, no guiding hand of purpose, but, holy God, Charlie and I were meant to meet.

"Hello, this is Charlie Yarish." A thick Brooklyn accent.

"My name is Bren Smith, an oysterman in Stony Creek. I wanna explore growing seaweeds."

"Come see me." Just like that. This guy doesn't mess around.

This moment marked the beginning of my deep dive into the underwater world of plants. Knowing little to nothing about land-based vegetation, let alone seaweed, I was an unlikely pupil, but I had a sneaking suspicion that, just as on land, plants might be easier to grow than animals—even shellfish. The deeper I dove, the more I found. To this day, I still can't quite fathom the fact that there are thousands of edible plants in the sea, and only a handful are being farmed. Why hadn't I done this before?!

I visited Charlie's test site, located off of Bridgeport, which is run by high school kids from the aquaculture vocational school. There are five hundred students, half of them from poor neighborhoods in Bridgeport, the other half bused in from the suburbs. Students get introduced to a number of marine-related careers. If aquaculture isn't their thing, they can learn how to build a boat, sail a boat, or use biotechnology to come up with a product that will solve a global issue.

We headed out to the seaweed farm, located two miles off the coast. Charlie had packed the boat with grad students from the University of Connecticut, EPA officials, and a gaggle of international scientists. He knew every seaweed scientist on the globe, and was regularly hosting experts from Brazil, China, South Korea, Mexico, Israel. Many were trained at his lab. I was the only farmer, and clearly posing above my pay grade.

We pulled up to the farm. I was expecting a complex system with steel, pulleys, rafts, all of which would have been well out of reach for a trailer-based oysterman's DIY budget. Nope, it was as simple as can be: just horizontal ropes attached to a couple buoys. In fact, it looked surprisingly like my vertical shellfish model. When the kids pulled up the lines, long shimmering blades of kelp emerged. Magical. Immediately, a glimmer of hope: this was definitely something I could do. The kelp was a beautiful dark brown—a wall of plants pulled from the deep. Charlie, beaming with pride, leaned over: "That dark brown is nitrogen!" He had an infectious way of bringing science to life.

Charlie and I agreed to team up. Right away, my email inbox

began to fill up with scientific articles authored by Dr. Seaweed. I couldn't understand the majority of it, but I didn't need
to. My job was to grow good food and get it into people's bellies. Charlie was to become one of the most important mentors
in my life—indeed, a lifelong friend who would later attend my
wedding.

He was a unique marine organism. Unlike so many in the
aquaculture space, he never tried to cash in. His heart beats for
people like me—folks who just want to make a living on a living
planet. Maybe it's his Brooklyn roots, but Charlie's vision of
applied science is what the world needs: experts who have their
eyes on both knowledge and social need.

With seaweed now incorporated into my vertical farm design,
the next step was permitting. I went online to download the
application—and, in an instant, all my dreams turned into pipe
dreams. Turned out there wasn't even an application for the
type of farming I was doing. Shit, there wasn't legislation for
seaweed, or for growing multiple species using a vertical farming system. My only option was to fill out the same application
used for permitting sewage-treatment plants, and the instructions alone were forty-eight pages long. Sucks being first.

Ron Gautreau, my right-hand man for many years, was a life-
saver through this process. Ron is good people. He has been
with me since my earliest oystering days and knows his shit.
He's owned and worked on boats his whole life, is recognized
as one of the top sea kayakers in the country, and, as if all that
wasn't enough, he's a trained biologist experienced in marine
permitting. My secret weapon. If there is one thing I've learned
over the years, it's to build a good farm team.

Ron worked for weeks, making farm drawings, mapping eel-
grass impacts, navigation lanes. He compiled hundreds of pages
of documents, navigating the guts of bureaucracy: the Army

Corps of Engineers, DEEP, NOAA Fisheries. Ron knew the secret sauce required to please the regulators. When he asked me to review and approve drafts, I'd just nod in agreement, unable to follow along. I didn't tell him that when I submitted my first oyster farm permit an agency staff member famously rejected my hand-drawn gear diagram by throwing it on the table and saying, "This application looks like it was done by a six-year-old!"

Permitting agencies get a bad rap, but I quickly learned that, at a staff level, agencies often welcome innovation. Problem was that the staff risk getting into hot water with the public for supporting something new; maintaining the status quo doesn't raise any political hackles. So our job was to flip the risk—create public pressure so they could only get in trouble for failure to act. This meant launching a mini–political campaign to raise the profile of the farm.

I began inviting politicians, environmental groups, and chefs out to the farm. I honed my message. It became clear that the jobs angle was the most potent. Politicians and Big Green groups were eager to link environmental protection with job creation, because for decades the enemies of the environment had convinced workers—coal miners, fishermen, farmers, loggers—that they had to choose between protecting their jobs and protecting the environment.

Out on the water, we could show concretely that jobs versus the environment was a false choice—that there were millions of jobs to be had fighting climate change and protecting the planet. The nexus between jobs and the environment was a pathway for rebuilding the middle class; indeed, addressing environmental and food crises offered up the largest job creation project since World War II. When Charlie, several students from the UConn School of Law, and I began drafting legislation legalizing seaweed farming (what later became a campaign to #legalizetheotherweed), we called it the "Seaweed Jobs Bill." Mine was the only job, but it worked.

Despite the enthusiasm within the agencies, permitting was painful and expensive—all told, it probably cost me five thousand dollars. Later, we were able to drive down the cost significantly for new farmers—closer to seven hundred—but at the time, I had to hustle a lot of T-shirts and oysters to pay the fees. The only real resistance came from duck hunters, but I solved that by explaining that the farm would attract ducks and create a prime hunting ground, so this meant they could just float around, drink more, and still bag birds. A duck hunter's perfect holiday.

When the permits came through, Ron and I began pouring anchors and setting gear for the newly designed farm. Seeding oysters and clams was easy: clams in the mud, oysters in cages lowered down to the seafloor. I picked up scallop seed from my shellfish co-op and planted them in the ten-tiered baskets of the lantern nets, about fifty scallops per basket, then hung the nets on every other horizontal line.

But seeding and planting seaweed was foreign terrain to me, having almost no relation to shellfish farming. I just stood back and watched. Charlie's team dove to collect spores off the shoals of Black Ledge, on the border waters of Connecticut and Rhode Island. They searched and pulled kelp blades that had switched into the reproductive stage, which is apparent when a raised dark stripe appears along the center of the blade. They arrived back at the hatchery with deepwater plants hungry to reproduce.

The hatchery team used razor blades to cut slits into the dark chocolate reproductive tissue and released the kelp spores into a prepared saltwater solution. This solution was poured into canisters holding two-foot sections of PVC pipe, each wound with two hundred feet of two-millimeter string. Overnight, the spores settled onto the string; then they were moved to twenty-gallon tanks—just regular old fish tanks. Eight pipes to a tank, forty tanks, with chillers keeping the water temp at fifty-one degrees and circulators keeping the water moving. The spores grew out for five weeks, until each kelp seedling was two millimeters long and ready to be outplanted on the farm.

The physical act of seeding was easy; timing was hard. Kelp is a winter crop and, in the best-case scenario, should be planted when the water temperatures are in the range of forty-eight to fifty-six degrees Fahrenheit. If it goes in too early, the seed will be smothered by other seaweeds. If the waters get too cold before outplanting, the kelp can't soak up enough nutrients to get through the winter. Plus, if the air temps drop below thirty-two degrees on the day of deployment, the seed will freeze and die.

To outplant, all I needed to do was unravel the strings around the horizontal lines on the farm. Each line took about five to seven minutes. Then, over a few weeks, the kelp spores transferred from the seed string to the main line. That's it! The only job of a kelp farmer was to control the depth of the line to ensure that the plants were getting the right mix of sunlight and nutrients, which meant attaching or removing buoys to control the depth of the lines.

Then just let the plants grow. As I suspected, compared with oysters, or other types of aquaculture and land-based farming, this was lazy man's farming. All I needed to do, once the lines were seeded and positioned correctly, was to check the farm every two weeks or so, to make sure the gear was secure: no fraying of lines, and anchors still secure. This is the benefit of growing species you don't have to water, weed, or feed. They just grow!

I've burned enough brain cells that I have trouble holding memories, but one clings: that first moment I pulled *Mookie* up to the newly imagined Thimble Island Ocean Farm. From two dimensions to three. From monoculture to polyculture. It was now a lush, layered garden, with multiple crops growing closely together in the confines of the simplest of structures. There was so much that it wasn't—no fish, no pens, no pollution. Just simple lessons taken from working within the seas; four vertical layers of crops. Kelp, mussels, scallops, oysters, and clams. I had asked the ocean what to grow, and it had answered.

CLIMATE FARMING

After a long voyage of trial and error, I had finally built a model that comfortably broke fully from industrial aquaculture. My daily work was now farming to breathe life back into a twenty-acre patch of water, and I loved it: body aching at the end of the day; thirst for meaning quenched; proud to carry home food for the table. With the help of many, I was now fully morphed from the pillager into the restorative ocean farmer. I had found my way back to the water.

But my twenty acres are not enough. My farm is embedded in an ecosystem that is dying. Some days, I want to hide from the tick of the climate clock. As I write this, extreme droughts are shriveling farms from Australia to Utah. Food prices are rising, and within our lifetimes, agricultural yields could plummet by as much as 70 percent. Recently, two new studies published in *Proceedings of the National Academy of Sciences* predicted that food shocks and malnutrition are to become the new normal if global warming goes unchecked. Why? Just as the world is running out of farmable land and freshwater, the global population is exploding.

The planet demands that we act, both fast and at scale. All the doom and gloom around climate change can be hard to stomach, but we're screwed if we keep doing business as usual.

What's emerging is a new climate economy, whereby water and land shortages will steadily drive up the cost of land-based food production. Add in the rising costs of fertilizer and animal feeds, and we either have to farm on Mars or to grow massive amounts of sea veggies and shellfish. In other words, in the face of climate change—small is no longer beautiful. We need to think big, act quickly, and scale up solutions. The time for little rooftop Brooklyn bee farms is over. We need to go big or go home.

Our seas can save us. With oceans covering 70 percent of the earth's surface and the majority of the United States lying underwater, there is plenty of space to build a new food system from the bottom up and scale to the level needed to address the climate crisis.

Want to scale food production? Restorative ocean farms have the capacity to grow massive amounts of nutrient-rich food. Professor Ronald Osinga at Wageningen University in the Netherlands has calculated that a global network of sea-vegetable farms totaling 180,000 square kilometers—roughly the size of Washington State—could provide enough protein for the entire world population. A modest goal of five hundred twenty-acre farms would produce 120 million pounds of shellfish and in the range of eleven to forty million pounds of seaweeds, depending on the species grown—that's a lot of food.

Want to scale employment? According to the World Bank, building a network of seaweed farms covering less than 5 percent of American waters could generate up to fifty million new jobs. If the revival of the middle-class blue-collar economy is the goal, this could be an important piece of the puzzle. But, just as important, this scaling does not require massive vertically integrated operations.

Rather, the future of ocean agriculture is networks—hundreds of small-scale farms sharing infrastructure and innovation, not thousand-acre banana plantations at sea. For growers of the new blue-green economy, scale is about replication, which is driven by setting low barriers to entry. My farm requires only twenty acres, twenty thousand dollars, and a boat, and water leases are only fifty dollars an acre. This means, unlike on land, where the price of farmland is a deal breaker for many beginning farmers, people from all walks of life can join us in building the blue farming economy, and with one farm netting up to $90,000 to $120,000 a year, it means a ladder of access from the poor to the middle class. Our crops are currently mainly sold as food, but they can also be turned into fertilizer, animal feeds, and biofuel. When we run out of people to feed, we'll weave seaweeds into a range of other industries.

All this scaling comes with zero inputs—no fertilizer, no feed, no land—making it the most sustainable form of food production on the planet. Our crops offer up more than sustainability: they have the power to breathe life back into the planet. Considered the "tree" of coastal ecosystems, seaweed uses photosynthesis to pull carbon from the atmosphere and the water; some varieties are capable of absorbing five times more carbon dioxide than land-based plants. And seaweed is one of the fastest-growing plants in the world. This turbocharged growth cycle enables farmers to scale up their carbon sequestration quickly.

Oysters, mussels, clams, and scallops also absorb carbon, but their real contribution is filtering nitrogen out of the water column. The main nitrogen polluter is agricultural fertilizer runoff. All told, the production of synthetic fertilizers and pesticides contributes more than one trillion pounds of greenhouse gas

EAT LIKE A FISH

emissions to the atmosphere globally each year. Much of this nitrogen from fertilizers ends up in our oceans, where nitrogen is now 50 percent above normal levels. According to the journal *Science,* excess nitrogen "depletes essential oxygen levels in the water and has significant effects on climate, food production, and ecosystems all over the world."

That same World Bank study found that dedicating 5 percent of U.S. waters to farming would also absorb ten million tons of nitrogen and 135 million tons of carbon. And since the crops grown on ocean farms can be used as organic fertilizers and animal feeds, they have the potential to radically disrupt the environmental footprint of land-based agriculture. We'll dive much deeper into these other uses for seaweed later in the book.

It took me years, but as my farm grew, so did the vision. I see a future of reefs: hundreds of twenty-acre ocean farms dotting our shorelines, surrounded by conservation zones. I see a Napa Valley of ocean merroirs, producing ocean vegetables with distinct flavors in every region. I see farms that are climate farms, producing zero-input food while sequestering carbon and rebuilding marine ecosystems.

According to Sean Barrett, the visionary cofounder of Dock to Dish:

> If done right, this new generation of ocean farming is poised to become the most restorative form of food production on the planet. We need healthy food and jobs that protect rather than harm our climate and Earth. It is a key piece of the puzzle for building a regenerative future.

CHAPTER 12

Look Back to Swim Forward

I often get asked how I invented 3D ocean farming. I didn't; individual inspiration is a myth. Invention is stealing, borrowing, listening, collaboration. I ripped off a thousand years of global history; my only innovation is synthesizing, designing with, not against nature. I was driven by a refusal to leave the water in the face of failure, intent on making a living. I learned from land and listening to the oceans, and I came along just as our food system was getting pushed out to sea. Mainly, I just got lucky.

Earlier, I covered the history of finfish farming, but there is also a parallel history of restorative farming. Often the two are lumped together into a single narrative, but they require distinction. Restorative farming began with shellfish.

It began in North America. Five thousand years ago, coastal First Nations began farming the sea. From Alaska to Washington State, they built man-made intertidal terraces, essentially small rock walls built into soft-sediment beaches. These walls provide extra stability to the sediment in prime clam habitat, thus extending productivity and increasing shell size. There's even a modern group called the Clam Garden Network made up of academics, researchers, and students that tracks the ancient clam gardens in the Pacific Northwest.

Beginning around 500 B.C., Romans began cultivating oysters, depositing collected wild seed on beds of broken pottery on the seafloor. By the first century B.C., Sergius Orata had learned how to spawn oysters in captivity by building rock piles on the muddy bottoms, which kept the oysters from suffocating in the silt. He made a fortune from his discovery; the Romans were the original foodies and obsessed with oysters. Emperor Vitellius, who lived in the first century A.D., was known to eat as many as a thousand in a sitting! Elite Romans also kept shellfish in vivariums, in-home ponds where seafood could be kept alive until dinnertime, an idea they stole from the ancient Assyrians.

Medieval Europeans continued Orata's practices, collecting and transplanting wild oysters for controlled cultivation, and little changed in this method right into the nineteenth century, when Europeans began scaling up their production. The problem was, before large-scale cultivation could take place, Europeans had to figure out how to collect oyster seed when these were just tiny swimming spawn. The Japanese, Chinese, and Mesoamericans in coastal Mexico had already been collecting seed for centuries—using woven bamboo and tree branches to collect the tiny seed oysters. By the sixteenth century, the Japanese had become adept at managing oysters and clams in captivity, using bamboo poles stuck upright in the mud and sand, a system still being used in Japan today. In France, Professor Jean Jacques Coste came up with a similar method in 1861, using bundles of twigs to collect oyster spat, and revitalized the dying oyster beds in France. Emperor Napoleon III took notice, and created two imperial oyster parks in the Bay of Arcachon. This area remains the center of oyster production in France.

Mussel farming was invented in the thirteenth century, pioneered, according to lore, by an Irishman who was shipwrecked in the Bay of Aiguillon, off the coast of France. Desperate for food, he tied nets to tall stakes to catch birds, but they were quickly covered by mussel seed. He found that the mussels would grow quickly after attaching to the nets, so he grew mussels instead. This practice became immensely popular, and it's the reason why there's a bay on the west coast of France known as the "Bay of the Stick."

At some point in the mid-nineteenth century, mussel farms in Prince Edward Island, New Zealand, and Spain began popping up, but much of this history has not been documented. How has so much knowledge of restorative farming innovation been lost at sea? In 1982, a Chinese aquaculture scientist collected bay scallops off the coast of Connecticut; of the couple hundred he managed to transport to China, only twenty-six survived, but they spawned the massive scallop industry now valued at half a billion dollars. Most scallops eaten in the United States today come from this Asian market, the offspring of those original New England scallops.

In North America, oysters had been harvested for popular consumption as early as the seventeenth century, but wild populations were rapidly depleted after the colonization of North America by the French, English, and Dutch, because of over-exploitation and the degradation of natural habitats. By 1955, the oyster industry was in crisis. One man is credited with the revival of oyster culture in the States: a Russian immigrant named Victor Loosanoff. In 1931, the U.S. Bureau of Commercial Fisheries built him a state-of-the-art laboratory in Milford,

Connecticut, where he went on to contribute countless innovations to the industry, most notably the first effective techniques for growing seed out to the "settlement stage" in a hatchery. In just fifteen years, growers in Long Island Sound were able to produce shellfish numbers comparable to their historical peaks. Techniques from the Milford Lab would go on to revolutionize shellfish production around the globe.

Today, the U.S. production of clams, oysters, and mussels tops more than forty million pounds a year and is valued at three hundred million dollars. Leading the way in mussel production are American Mussel Harvesters in Rhode Island and Bang's Island Mussels in Maine. California's Hog Island Oyster Co., which started as a five-acre lease in 1983, now grows three and a half million oysters a year. Taylor Shellfish is the country's largest shellfish producer, harvesting more than two million pounds of shellfish annually and has done the most innovative work on shellfish in North America. In the little state of Rhode Island alone, oyster growers raked in more than $4.3 million in 2016.

In addition to these major producers, small oyster farmers blossomed around the country. Bob Rheault, executive director of the East Coast Shellfish Growers Association, cites massive growth over the past two decades. "You've got 1,000 small farms or more than that on the East Coast alone," Rheault said. He estimates there are three to four hundred separate shellfish brands now, each from a different farmer. In the Chesapeake Bay, oyster production skyrocketed by 806 percent from 2006 to 2012, mainly from small producers.

In Galicia, Spain, farmers have taken mussel farming to massive scale. This is one of the largest mussel-farming areas in the world, with world-renowned quality. Galician mussels are farmed using floating rafts, and there are more than three thousand of these platforms spread throughout the region. With its production level of over two hundred thousand metric tons, the Galician mussel industry has generated more than eight thousand jobs and incorporates a thousand farming support vessels.

GROWING SEA GREENS

The first mention of seaweed farming appears in Korean history texts in the fifteenth century, and there are records of farms in Japan's Tokyo Bay as early as 1670. Today, more than six and a half million tons of sea vegetables are farmed in Asia each year, and it's an industry branded as an Asian invention. But, in fact, the history of how this industry grew is a more complex tale of innovation and collaboration between the East and the West.

The modern Japanese seaweed industry is credited to the pioneering work of the British seaweed scientist Kathleen Drew-Baker. Traditionally, Japanese farmers threw bamboo branches into shallow, muddy water, where the spores of the seaweed would collect. A few weeks later, these branches were moved to a river estuary. Later, farmers deployed nets strung between the poles to wait for nori seed to build up on them. This method was used for hundreds of years but remained small-scale, with the vast majority of seaweed sourced from the wild.

Sourcing seed like this made farming unscalable. Wild collection was the luck of the draw. Some years, large quantities of the thin filament-like spores grew into healthy, harvestable plants with long, green leaves; other years, they failed to settle. Since little was known about nori's life cycle, there was no way to spore new seed to repopulate the depleted seaweed beds.

Wild sporing became increasingly problematic as industrial-

ization polluted waters and a string of typhoons led to a disas-
trous drop in harvests. By the late 1940s, nori production in
Japan had dwindled to almost nothing.

But then up popped Dr. Kathleen Drew-Baker in Manchester, England, who had been a lecturer in botany at the University of Manchester, where she studied algae. The university
barred married women from lecturing, so, when Dr. Baker tied
the knot with fellow academic Henry Wright-Baker in 1928, she
was fired and relegated to a job as an unpaid research fellow.
Undeterred, Dr. Drew-Baker and her husband built a seaside
lab, and here she dedicated herself to figuring out how to reproduce algae spores. She focused on a type of nori known as laver
that grew off the coast of Wales and was used locally in bread
and soup.

In 1949, she published a paper in the journal *Nature* detailing
her work, and the timing was perfect. Back in Japan, Sokichi
Segawa at the Shimoda Marine Biological Station read Drew-
Baker's paper and combined her research with new techniques
of using synthetic material tied to bamboo poles. This practice
rescued the nori industry and is still used today by farmers
like me.

Though Dr. Drew-Baker is all but forgotten in the United
States and Europe, in Japan she is a hero and has been named
"Mother of the Sea." In Osaka, there is a monument in her
honor; in Kumamoto, a shrine. Every year since 1953, there has
been a "Drew festival" to celebrate her legacy.

A similarly untold tale is that, during the same time the
Japanese were scaling their operations, U.S. scientists and engineers were launching a major effort to jump-start a domestic
seaweed-farming industry. But the two countries had radically
different goals. Unlike their Asian counterparts, who farmed for
food, scientists in the United States wanted to farm seaweed as

an industrial ingredient. As far as I can tell, this decision, driven by differing national tastes, is why seaweed farming failed miserably in U.S. waters.

As far back as the eighteenth century, coastal economies in Europe and the United States had harvested wild kelp to turn into a potassium-rich ash for pottery, glass, soap, and textiles. Entire regions were transformed into industrial production centers. According to Claire Eamer, a journalist who has dug deep into the boom-and-bust history of Western kelp production:

> When the Napoleonic Wars cut British factories off from continental kelp suppliers, they turned to the seaweed-rich islands of Scotland. Island economies went into overdrive, and the lairds, who owned the land, made fortunes. But the war ended, and British factories turned to cheaper continental kelp. All that remains are a few shallow, circular depressions—abandoned burning pits. When the kelp money left North Ronaldsay, so did whole families. Like other Scots, some probably washed up on the far side of the ocean in Atlantic Canada. If so, they were on the spot for another North Atlantic seaweed tsunami—the Irish moss boom of the 1950s and 1960s.

Rather than being eaten as a sea vegetable, Irish moss, which is 40 percent carrageenan, was turned into a thickener or stabilizer for processed foods, ranging from chocolate milk to beer. Between 1950 and 1970, Canada became the world's leading supplier of carrageenan, allowing remote outposts to profit from the industrial revolution.

We even turned kelp into a weapon of war. During World War I, California's kelp beds were deforested for conversion into acetone and potash for munitions. Massive banks of kelp digesters lined the coast of Chula Vista, California, operated by the Dupont-owned munitions company Hercules Chemical Company, which processed kelp to extract ingredients for explosives. After the war, kelp continued to be heavily harvested for fertilizer and animal feed, as well as nutritional supplements. With World War II looming, the U.S. military became increasingly concerned that demand would outstrip supply, both from within the United States and from foreign sources in Japan. This triggered interest in whether seaweed farming could fill the anticipated gap.

Along came a University of Michigan graduate student, C. K. Tseng, who defended his phycology (marine botany) dissertation five months after the attack on Pearl Harbor. Tseng, funded by the U.S. Department of Agriculture, started to explore alternative sources for a range of seaweed extracts. Working from the Scripps Institution of Oceanography, Tseng conducted groundbreaking work in algae reproduction, cultivation, and processing, and eventually concluded that Pacific waters were well suited to farming. But just as he began plans to build a test farm, his funding dried up and he decided to return home to China. It would be more than thirty years before a new generation picked up where he left off. Tseng's work is credited as a keystone to the later growth of China's mariculture industry. In fact, in 1983, when California's kelp beds were decimated by El Niño, San Diego–based Kelco began importing Tseng's Chinese *laminaria* as raw material to repopulate the waters.

The next wave of U.S. interest in ocean farming was spurred by the 1970s energy crisis—again, not for food, but for fuel. The National Science Foundation and U.S. military funded a research team led by Dr. Howard A. Wilcox, an engineer who made his reputation with his work on the deadly sidewinder missile. He advocated for massive kelp farms along U.S. coastal waters and sketched out designs. He assembled a team at the Naval Undersea Center in San Diego to explore the viability of his new idea: oceanic farming. The project description was ambitious. In a 1973 paper titled "Concerning the Selection of Seaweeds Suitable for Mass Cultivation in a Number of Large, Open-Ocean, Solar Energy Facilities ('Marine Farms') in Order to Provide a Source of Organic Matter for Conversion to Food, Synthetic Fuels, and Electric Energy," issued by the U.S. Naval Weapons Center, two of his colleagues proposed:

> [A] thousand-acre proof-of-concept kelp farm . . . would be built in 1982, produce about 400,000 tons of kelp per year for conversion, on site, to fuels, petrochemicals products, and feedstocks for animals. Wave-actuated pump would bring nutrient-rich cold water up from 1000-foot depth. Additional food for giant kelp plants transplanted to submerged mesh would be distributed through farm's structural members.

An illustration of the proposed "Hemidome" was published in *Popular Science* in July 1975. It's insane: massive propellers, huge steel cables, thousands of lines of kelp surrounding a floating platform housing storage and processing machinery akin to an oil platform. It looks industrial and monstrous.

Meanwhile, scientists like Dr. Michael Neushul, a dynamic and creative marine biologist out of the University of California in Santa Barbara (UCSB), began trying to grow kelp off the coast of Goleta, California. He founded Neushul Mariculture Inc., seeded with funding from the Department of Energy, which saw kelp as a path to oil independence because of its potential use for biofuel production. Neushul's first design was modest, based on attaching giant kelp seedlings to eighty-pound bags of gravel. He actually built and outplanted the farm—the first in the United States—with his "gravel bag method" off the Elwood piers. It was experimental and innovative, a family affair. His sons were the primary research assistants—and cheap labor.

Neushul's installation may have been small-scale in practice, but the stated goal of his research was much more grand, with a focus on industrial production—fuel, chemicals, and feedstock—rather than food. The project was housed under the Department of Energy rather than the Department of Agriculture, and associated with large corporations like General Electric. This aroused skepticism in the press, and kelp farming became the poster child for wasted taxpayer dollars.

In February 1980, *The Washington Post* ran an op-ed by Jerry Knight accusing General Electric of blowing through $1.2 million of taxpayer money to research seaweed biofuels. Knight called it the "kelp calamity" because:

> General Electric in December of 1978 carefully transplanted 100 kelp plants onto a quarter acre of Pacific Ocean floor. Within two months, all the kelp was gone. According to an internal report assessing the "exotic energy project," the seaweed scientists not only had trouble keeping track of their kelp, they also apparently got their money for the project from the government when private funds were readily available.

A cartoon accompanied the op-ed, showing a DOE bureaucrat inspecting a kelp farm marked with a sign "U.S. Govt. No Trespassing Kelp Ranch" and yelling into what looks to be a giant cell phone, "Kelpnappers!" When I look back at the cartoon, what's interesting—besides its surely being one of the first artifacts of kelp satire—is that it signals an early fear of corporate-led industrialization of the seas. This foreshadows the permitting wars that were on aquaculture's horizon.

Mother Nature ended the project abruptly in 1983, when El Niño destroyed Neushul's farm.

In 2018, I had the honor of meeting Michael Neushul's son, Peter Neushul. Appropriately enough, it was at a 2018 Department of Energy gathering on kelp farming, where Peter presented an engaging lecture on the history of ocean farming. Looking back on his father's work, he was proud, but also reflective on the failures. He criticized the use of slow-release fertilizers attached to the lines, which replicated problematic methods from land-based agriculture. He also reflected on past constructions such as the Hemidome, designed largely as a plastic array, which he said was overly complex and doomed from the start. It was thrilling to learn from him, and to gain an understanding of my place in the kelp farming lineage.

We're standing on the shoulders of seaweed pioneers like Dr. Neushul and my mentor, Dr. Yarish. In 2014, I toured Dr. Neushul's abandoned research facility. It was like touring a dead tributary of history. Test tubes, diving suits, water tables, greenhouse, grow lights, buoys, a shipping container of research. Dust, weeds, broken glass. An abandoned kelp playground. I wondered what Neushul would think if he saw our thriving ocean farms now. Looking forward, his son Peter is hopeful: "We're a Third World country when it comes to mariculture, but we can be a paradise for ocean farming."

Seeding Season

The permit's in hand, your farm's built—so now it's time to seed the shellfish and kelp. When I shifted from fisherman to ocean farmer, I ran a death camp my first few years, killing literally millions of baby shellfish while I figured out what the hell I was doing. We're not going to let that happen to you.

Shellfish and seaweed hatcheries are scattered along the coasts; you just need to know how and where to look. Some are private; others are housed at universities. Your local NOAA Sea Grant office will likely have a list. When you are buying seed from a hatchery, make sure that you are selecting the best brood stock available. You want seed with a proven record of good growth and survival in your particular region and body of water.

You should be planting seed in each of the four seasons.

Kelp and mussels can go in during late fall or early winter. For oysters, clams, and scallops, aim for early spring. Oysters can go in during the summer, too.

Let's start with kelp seed. It will be sold as eighteen-inch spools wrapped with seed string. Each year, the hatchery staff gather a few kelp blades from the wild and induce sporing in tanks filled with filtered seawater. They wrap 250 feet of specialized nylon string—essentially, a purer formulation of household twine—around the PVC spools, which are put in the tanks with spores. Within twenty-four hours, the spores attach to the string. Over the next four to six weeks, the kelp will grow out to one or two millimeters in length. Each spool will cost you around $150, depending on the hatchery.

It's important to seed under the right conditions. Kelp spores thrive after water temps drop to between fifty-six to forty-six degrees Fahrenheit, which here in Southern New England is typically October or November. On the day of seeding, air temperatures need to be above freezing, factoring in windchill—I aim for forty degrees or above to be safe. Many a farmer has lost an entire crop because of setting out seed on too cold a day.

The process of outplanting is surprisingly easy. Untie one end of your horizontal longline, and feed it through the seed spool. Unravel two feet of seed string, and splice it into the line, using your fid (as an ocean farmer, you're gonna become a master splicer). Then slowly motor your skiff down along the longline, allowing the seed spool to unravel around the line. Your job is to make sure it unwinds evenly. Be prepared to stop quickly if you develop a tangle in the seed string. Every fifty feet or so, stop the skiff and splice a float with a five-foot tail onto your longline. When you unravel the entire spool, splice the end of the seed line into the longline, again using your fid. You're done. Each line should take five to ten minutes to seed, so you can easily seed in under two hours, even with the usual hiccups. Over the coming weeks, the spores will transfer themselves from the string to the longline and begin growing vertically downward.

To seed scallops, first purchase seed from a hatchery for direct outplanting into lantern nets, which are ten-tiered mesh columns. Collecting wild spat is unreliable and not recommended. For bay scallops, aim to buy seed at eight to twelve millimeters or larger (up to twenty millimeters, though the larger seed will be more expensive). Sea scallops should be purchased at about five to ten millimeters, when they are about a year old. Make sure that the seed is larger than the mesh of your lantern net. I recommend that you always err on the side of larger seed when possible: you will have significantly fewer mortalities.

Besides kelp, mussels are my favorite crop, because they grow faster than other shellfish, are nutritionally superior, and do not require a hatchery phase, which makes them less expensive to grow. I actually began farming mussels by mistake. One year, I was lazy and left my kelp lines in the water after harvest season. After kelp harvest, the lines are covered with the kelp "roots" where the stipes (stems) attached to the lines. It turns out that mussel seed loves to attach to these roots. Every spring, right after kelp harvest, mussel seed floats through the water column and sets on the old kelp lines. Repurposing the same gear for different uses throughout the year is efficient and drives down costs.

Over the next two months or so, the mussels will grow out to one inch or more. Once they reach adequate size, pull the longlines aboard your boat, and strip off the mature seed into buckets or fish totes. Take your mussel socks, cut them to lengths of fifteen feet, and begin filling them with seed. It's like stuffing sausage. I usually do it by hand, using a PVC pipe on the inside of the mussel socks, but some years I use a socking machine, which churns out 250-foot-long continuous socks that are strung in loops along the longlines. Otherwise, I tie the fifteen-foot mussel socks every three feet along the longlines, weighted at the ends with a few pounds of chain link, or whatever scraps of metal I have lying around. This keeps the socks from floating up to the surface.

Oyster seed is available from a variety of places and in a variety of life stages. You can get any size of oyster seed you want; smaller seed is cheaper but has a much higher mortality rate. Quality hatcheries carefully select for shape, size, and meat content. To grow oyster seed, they take full-grown oysters, put them in tanks, and raise the water temperatures to force them to release millions of their invisible larvae. These spawn are fed red, green, and yellow algae, while salinity, nutrient level, and other environmental requirements are carefully maintained in the tanks. For the small-scale backyard grower, I highly recommend purchasing mature seed that is at least one inch in length. This will prevent the need to change the size of your mesh containment systems as the oysters grow, and will limit mortality.

Oyster culturing allows you to plant two seedings each year, so you can have a more consistent supply. Since oysters will be growing fastest in the summer, they will need more attention during those weeks, as you harvest the biggest ones and remove biofouling. Think of biofoulers as the weeds of the ocean—everything that's growing that you don't want, including algae, barnacles, and sea squirts, to name a few. Oysters in bottom culture take eighteen months or so to reach full size. Oysters in lantern nets and floating bags and cages will mature more quickly, and may be ready between six and eighteen months. Growth will vary a lot, depending on temperature, salinity, food availability, height in the water column, and population density. You just need to keep an eye on your oysters and harvest the largest ones periodically. During the coldest months, when your oysters are in hibernation, you should leave them undisturbed.

When you're choosing clam seed, I recommend purchasing larger seed to minimize your grow-out time and risk of predation. It will be more expensive at the start, but will pay off in survival rates and time saved. Seed for grow-out will usually range from ten to twenty millimeters, and you should look for seed that is in the upper end of this range. Make sure you know when the seed you are purchasing spawned, to be sure you aren't buying slow growers. It will generally be shipped to you in bags inside coolers, with ice packs, and should remain cool and moist, not submerged in water, until outplanting. Outplanting is easy: just shovel the seed into the water between your lines of kelp, and let it settle into the mud on the ocean floor.

part **4**

CHAPTER 13

Falling for a Foodie

In 2013, I fell in love through my CSF. One of my members invited me over to dinner to meet his wife and kid. I showed up, and had been hanging around awkwardly for a while when in walked Tamanna Rahman. She was loud and warm—and beautiful. Months later, my mom characterized her as "dangerously smart." She was studying nursing at Yale, a first-generation Bangladeshi raised in southern California, a former union organizer at West Virginia casinos, and a consummate foodie. She read cookbooks in bed at night and could draw you a map of taco trucks in L.A.

After being admitted to graduate school in Connecticut, she wrote up a list of five things to do when she arrived in New Haven. Top of the list was joining the New Haven shellfish CSF, which she'd read about in *Lucky Peach* magazine. After the dinner party, I tried to woo her by sending over a box of my shellfish. A few hours later, she texted me a photo of my clams swimming in a broth with lemongrass, cilantro, and chilis. A few hours after that, I got another text: she had broken out in hives. It was her first allergic reaction to shellfish, and this unfortunately would only worsen over time.

We spent weeks after that texting like tweens, thousands of texts that I would one day print out and paste into a journal as a birthday gift for her. It was over a month before we saw each other again. On our first date, she took me to a lecture by Harold McGee, who writes about the science and history of food. She was insistent we had to be early, convinced there'd be lines out the door for such a celebrity. When we arrived, the auditorium was empty. But for her, McGee was a rock star. During the lecture, she sat at the edge of her seat, clutching her dog-eared copy of his book.

Afterward, we spent hours wandering the streets, chatting easily about our work lives, our families, our history, our dreams. She asked me if I wanted to see the route she walked to the hospital. It was an odd suggestion for a date. She stopped to show me buildings she liked, where she bought donuts, her favorite graffiti tags en route. That walk nipped my heart. Showing me her daily commute, Tamanna was letting me in on what made her who she was: pride in work. It was at that moment that I understood how, despite coming from deeply different origins—her urban, me rural; her brown, me white; her refined, me uncouth—we shared a foundation built with the bricks of meaningful work and everyday street-level lyricism. We've been doing those walks now for years, nearly every evening, and never run out of things to talk about.

Tamanna pushed me into new prisms. Food for sure, but also gentler ideas that were foreign to me, a waterman. Early on, she presented me with a test. She casually gave me a favorite novel to read, *The End of the Affair* by Graham Greene, and mentioned that one of her favorite lines in all of literature was in that book. She teasingly dared me to find the sentence—but I knew the stakes were high. It was 160 pages long! I read and searched. It was about two people illicitly falling in love. Much of it I found boring—I had been reading adventure stories like Farley Mowat's *The Boat Who Wouldn't Float* and Steven Callahan's *Adrift: Seventy-Six Days Lost at Sea*. But I was falling hard for her, so I picked through each page.

A third of the way through, the main character, Maurice Bendrix, telephones a woman to ask her to dinner. For years, he'd suffered self-hatred and hidden his love. But now she agrees to see him. After hanging up the phone, he writes, "And sitting there, my fingers on the quiet instrument, with something to look forward to, I thought to myself: I remember. This is what hope feels like."

This is what hope feels like. I had found it. When I presented my discovery, Tamanna looked at me suspiciously, as if I'd somehow cheated and peered into her head. She later said that it was the moment she knew she would marry me.

Less than a year later, I proposed to her. I borrowed a friend's rusty barge and set it up for a floating dinner on the farm. Chairs, table, white linen tablecloth, candles, champagne glasses. For the meal, I made grilled cheese sandwiches and tomato soup on a Coleman grill. I know, doesn't sound romantic. But when I was a kid, my family had a tradition of cooking grilled cheese on the beaches of Newfoundland. For a hick, there was romance to it, the type of romance that I suspected Tamanna would appreciate.

I knew how to make a grilled cheese, but not for a girl who expects seven types of cheese and homemade bread, so I called in a favor with Jason Sobocinski, a chef in New Haven who had a show on the Cooking Channel all about cheese. He hooked

me up. Gave me a tutorial and sent me packing with the ingredients, too: big hunks of hours-old fluffy bread, handpicked
mix of blended cheeses, creamy local butter. He even supplied
the tinfoil and a brick to weigh the sandwich during grilling.
I brought Campbell's as the tomato soup, because sometimes
you just gotta stick with the classic.

Everything was in order, but there was a flaw in my plan.
Tamanna gets seasick and is lactose-intolerant, so before I fer-
ried her out to the barge, I had her take a Dramamine and some
lactose pills. Once we arrived at the farm, we had a glass of
champagne and then I popped the question. She jumped up,
shocked and dazed, almost falling into the water. I steadied her,
and asked again: "Will you marry me?" She said yes, but she was
so woozy from Dramamine and booze that I asked her again the
next day to be sure. We cooked and ate, laughed, talked about
the life we would build. It was a good day on the water. Two
things I loved coming together: my farm and nurse.

Turns out there is a long tradition of farmers marrying
nurses. In fact, during our honeymoon, a Danish farmer we met
told us that agricultural colleges were often built next to nursing schools for this very reason. He shared a saying: "To make a
living, every farmer needs a nurse as a wife—and some seasons,
he needs two." Tamanna told me not to get any ideas.

In the beginning of our relationship, we spent more time
with my mom than I had in decades. She was living a couple
hours away, in Rhode Island, with her partner of thirty years,
Sylvia. Tamanna was always homesick for her family in Los
Angeles, and she and my mom hit it off right away. We'd all
spend weekends together on their couch, eating, drinking, and
watching Netflix. We adopted my mom's habit of turning any
occasion into a celebration, and our daily lives felt richer for
it. When she died suddenly, a year later, it felt as if a new sem-

blance of family we had just started building came sharply to an end. I still can't look at pictures of her.

Soon after, Tamanna and I bought the house of a nineteenth-century East Coast oyster captain. His name was John B. Ludington, and, according to lore, he was the inventor of deepwater oyster cultivation. There is a shucking room in the basement, and a thick bed of oyster shells two feet below my lawn. From our bedroom, I can watch oyster boats heading out for a day's work at dawn.

I finally felt rooted for the first time since I left the banks of Newfoundland.

CHAPTER 14

Center of the Plate

Native seaweeds contain more vitamin C than orange juice, more calcium than milk, and more protein than soybeans. Those on the hunt for omega-3's are often surprised to learn that fish don't create these heart-healthy nutrients by themselves—they consume them. By eating the plants fish eat, we get the same benefits, while reducing pressure on fish stocks. So it's high time that we eat like fish.

Problem is, a lot of Americans think seaweed is disgusting. For most of my life, so did I. Slimy and rubbery, it's got "weed" in its name, and it's more likely to wash up on the beach than show up on your dinner plate. Sure, dried seaweed snacks and seaweed salads are widely accepted now. Even Costco stocks seaweed these days, and you can find California rolls at your local gas station. But if we want to leverage the thousands of edible ocean plants to build a new agricultural food system out to sea, snacks, side dishes, and the occasional sushi dinner aren't going to cut it. We've got to move ocean veggies to the center of the plate, and that means convincing Americans that sea vegetables are no different from any other vegetable.

Unfortunately, I'm a terrible choice for the job. Despite spending most of my life catching or growing food, I'd never really learned to talk the foodie talk. One time, at a party on a friend's land-based farm, everyone had to go around the table and say their favorite vegetable—these are the kinds of situations my wife gets me into. I became the laughingstock with my answer: bread. I mean, grain grows from the dirt, so it's got to be a fruit or vegetable, right? I've been dumping toxic industrial slop into my body for decades, and I still can't identify most of the vegetables we get each week in our CSA. I'm forty-five now, and with the way I've eaten, I'll be lucky if I have ten years left.

My wife, on the other hand, grew up in Los Angeles, home to the largest population of Asian Americans in the United States. She was munching on kimbap (think Korean sushi rolls) and slurping down seaweed soups long before I ever imagined I'd be trying to eat the slimy stuff on my fishing nets. When she moved to the East Coast for college, out of range of her L.A. cuisine, she adapted by becoming a fantastic home cook. If anyone was going to be able to help me learn to love the sea veggies I was growing, I knew she'd be the one. Together we began learning about ocean vegetables.

We're not the only ones on the hunt for new cuisines that will feed the planet without destroying it. Some are breeding bugs; others are growing artificial meats. One time, I was on a panel on the future of food at the Boston Museum of Science. Next to me was a woman from NASA who was working on the "Menu for Mars." She was super-engaging to listen to, but I was stunned by the idea this was our only hope now: We have to go to Mars to eat? I'm convinced that our oceans can provide more than enough delicious food right here on this planet.

The stakes are high. If we want to keep eating delicious food, we need our top chefs to figure out how to make climate cuisine into haute cuisine. And how exciting a proposition! Imagine being a chef in 2019 and discovering that there are thousands of vegetables you've never cooked with. It's like discovering corn, arugula, tomatoes, and lettuce for the first time. As the chef Brooks Headley, who would later play a key role in unlocking the culinary power of sea greens, said when I first gave him kelp to play with, "As a chef, it feels frightening, daunting, and exciting all at once."

A LOST CULINARY HISTORY

The deeper I dove into seaweed history, the more I realized that creating a Western market for sea vegetables wasn't such an outrageous proposition. In fact, there are numerous untold culinary and agricultural traditions of seaweed use far beyond East Asia. Anywhere there are oceans, people have had a relationship to seaweed, including in the kitchen.

In the archaeological community, there is something known as the "Kelp Highway." In 2008, remnants of nine different seaweed species dating back to around 12,000 B.C. were found on hearths and tools in Monte Verde, a site in Chile. These included varieties of *Gracilaria, Porphyra, Sargassum,* and *Durvillaea antarctica.* This discovery provided new evidence that seaweeds were important in the diets of early inhabitants of the New World, and were frequently cut and prepared with stone tools. Some bits of algae were found burned, suggesting they were dried for storage and transport or were cooked. Seaweed remnants were also found in the form of masticated cuds, lending further evidence that they were an important part of the diet. This settlement is located hundreds of miles from the coastline, indicating that the seaweeds were gathered from the coast and transported inland for year-round sustenance.

Seaweed was also part of the diet of the Incas, the empire that ruled much of current-day Peru from the thirteenth to the sixteenth centuries. The Incas ate a variety of sea vegetables, all called *cochayuyo* in Quechua, from marine *Porphyra, Gigartina,* and *Ulva lactuca* to the freshwater cyanobacteria *Nostoc*. Some were eaten fresh, and some dried for preservation or trading. After the Spanish conquest in the mid-sixteenth century, written records kept by conquistadores note that *Nostoc* was often used to make a dessert by boiling it with sugar and water.

As we also know from the writings of Spanish conquistadores, pre-conquest Aztec people living in what is now Mexico City also feasted on algae, such as *tecuitlatl,* which is likely what we know today as the common dietary supplement spirulina. The *tecuitlatl* was collected with fine-mesh nets and spread out to dry in the sun, then pressed into loaves and cut into bricks. The flavor was described as similar to cheese in texture and flavor, so much so that the Spaniards called it *queso de indio,* translated to "Indian cheese." *Tecuitlatl,* which was very salty and blue in color, was eaten with toasted maize or tortillas, and sometimes used to make a sort of tortilla itself. It was reportedly widely eaten among the Aztecs; the saltiness of the dried algae preserved it for up to a year, allowing it to be traded and sold in faraway areas outside of the city.

Lending further support to the Kelp Highway hypothesis, some of the earliest coastal settlements that have been found are located near productive kelp forests, and remains of varied seaweed species have been discovered in middens along the proposed route of coastal migration.

It's well known that seaweeds made up an important part of

the diet for the First Nations of the Pacific Northwest, who ate a variety of red and brown algae and traded them with inland peoples for other goods. Seaweeds were generally harvested at low tide and preserved in one of a variety of ways. Some were dried on stones and broken into pieces for storage in wooden boxes, to be eaten later, either alone or cooked in fat with fish or shellfish. Some were roasted and smoked over open flames, then powdered to be mixed with water and whipped into a dessert. Some seaweed was partially dried, then fermented in a wooden chest layered between planks of cedar for flavor, weighted down with a stone for about a month. The final product was a flat seaweed cake that withstood long-term storage and was eaten chopped and cooked with fat or in soup.

Seaweed has also long played a role in diets across the Atlantic. According to the Scottish marine ecologist Iona Campbell, traces of ocean vegetables have been found in Orkney Island cremation sites dating back to the Bronze Age. Irish monks in the twelfth century reportedly gathered dulse (dillisk) along the shoreline to give to the poor. The Irish language is disappearing, and with it the thirty-one Irish names for seaweed. *A Celtic Psaltery,* from the sixth century in the Hebrides, reads:

> *What greater joy could be?*
> *Now plucking dulse upon the rocky shore*
> *Now fishing eager on*
> *Now furnishing food unto the famished poor.*

I came across a more recent—nineteenth- or early-twentieth-century—Irish saying that women should be concerned with "Potatoes, Children, Seaweed." The Irish also used seaweed medicinally, as a cure for colds, worms, and women's homesickness.

Farmers, fishers, and livestock raisers were granted the right to harvest seaweeds from the coastline to augment their income for many hundreds of years. During times of poverty, dulse was incorporated into bread flour in order to stretch the dough a bit further. In the twentieth century, toasted dulse was commonly served as a snack in Scottish and Irish pubs in order to induce salivation and to encourage patrons to drink more beer. Carrageen moss, or Irish moss, so-called for its widespread use in Ireland, has been historically used as a thickening agent in jellies and puddings. It was used as a folk medicine for colds because of its high iron content and nutrition.

Ever heard of the Maritime and Irish Mossing Museum in Scituate, Massachusetts? I used to lobster right off the shore, but I never heard of a seaweed museum until I started drinking the Kool-Aid. The museum displays the tools of the backbreaking trade, including dories, creels, rakes, and floating oil dispensers. The star of the exhibit is the Irish fisherman Daniel Ward, deemed the father of Irish mossing. According to local lore, he visited Scituate in 1847, looked down at the rocks at low tide, and recognized the red algae from back home: Irish moss. He founded a successful company and started hauling ashore up to two thousand pounds a day to use as a thickener and stabilizer in beer, chocolate, pudding, and other foods.

Over in Wales, laverbread, or purple laver, is a spinachlike purée of boiled seaweed that has been eaten for breakfast since at least the seventeenth century, traditionally heated in bacon fat and served with clams. It is also eaten in salads, biscuits, and with roasted meat. During the time of the industrial revolution, it was gathered by the wives of coal miners, who cooked and drained it before selling it at market. In the eighteenth century, English whalers often dried laver and ate it during long sea voyages to help stave off scurvy. Though its popularity is declining today, laverbread is still widely eaten by the elderly, affectionately referred to as "Welsh caviar" for its texture.

Seaweed has also been integrated into the hyperlocal diet of many southern Italian towns and islands, where it is a readily available ingredient to add nutrition and flavor to a dish. In Catania, *u mauru* is the name for red seaweed that is traditionally eaten raw with a little lemon juice. Variations on this preparation include soaking it in wine vinegar for half an hour before draining and sautéing with garlic and olive oil. *U mauru* has mostly disappeared from the cuisine of today, because the seaweed is no longer abundant in the coastal waters. In other towns of southern Italy, seaweed is often incorporated into savory fritters called *frittelle di alghe*.

Icelandic people have been eating dulse, or *söl,* since at least 960. Consumption of the seaweed is mentioned in a story from the tenth-century saga of Egil Skallagrimsson, where it is described as a food that has some capability of restoring the vitality of the eater. The oldest Icelandic law book talks about the rights involved in the collection and harvest of the seaweed on a neighbor's land. Beginning in the eighth century, dulse was used as a currency and was often traded between coastal and inland populations. The dulse was eaten fresh baked in bread or dried for a snack. It was sometimes mixed with butter for flavor, or used as a thickening agent in milk and porridge. The consumption of dulse has decreased in recent years, partially because of its association as a poor man's food. "He was laid in his grave with a piece of seaweed in his mouth" is an expression in the Faroe Islands, where some of my ancestors are from, meaning that someone died poor.

In the United States, there have been efforts throughout the twentieth century to incorporate seaweeds in the American diet. Multiple newspaper clippings from the early years of the century urged Americans to adopt seaweed as a food staple. An article in *The Boston Cooking School Magazine of Culinary Science and Domestic Economics* from 1912 breathlessly told readers: "Just think of America, literally surrounded with all of these wonderful plants, in the cultivation of which lies many fortunes, and not realizing her boundless wealth in her marine forests!" A message I couldn't agree with more. Another article, from a 1918 issue of the *Hartford Courant,* reads, a bit less enthusiastically, "In our search for food substitutes we have come face to face with seaweed. We have not begun to put it into our faces to any extent as yet, but here it is right in front of us." As apt in 2018 as it was in 1918.

LEARNING TO LIKE KELP

After all this research, sea vegetables were starting to sound a whole lot more appealing. Cooked with bacon and clams, baked into scones, toasted like cheese on tortillas? Sign me up for the first sea greens quesadilla.

I was also encouraged by the existence of this secret history of ocean vegetables. It suggested that the journey we were on was rooted in something deep in the cultural fabric of our nation and the many peoples who make it up. We're in a historical moment in which people are hungry to rediscover and revive these historical food traditions. Take Sean Brock, an award-winning chef credited with recovering traditional Southern cuisine. Partnering with Anson Mills, the grain company, he has focused on resurrecting heritage varietals of grains and legumes, particularly on Carolina Gold rice, which was nearly extinct before Brock highlighted its importance to Southern foodways. Or Dan Barber, the acclaimed chef who has taken a deep dive into rediscovering delicious, forgotten varieties of squash, potatoes, grains. Just last year, he launched his own seed company, focused on developing vegetables the way they used to be before mass commodification—that is, growing what's delicious, not what can travel well and sit on the shelf the longest. I had a new vision for our sea greens.

Trouble was bringing this vision to fruition, and finding others who might share it. One of our first tries at getting kelp into the spotlight was back in 2012. I got in touch with Mayur Subbarao and Nora Sherman, who were minor stars of the local food-and-beverage scene of New York. A self-described cocktail geek, Mayur was known for weird drinks and what *The New York Times* described as "harvesting . . . more esteemed ice" throughout Brooklyn for his cocktails. He and his partner, Nora, were running pop-up cocktail events around the city, and decided to use kelp as one of their themes. They roped in the Portuguese chef Dave Santos, who was at the helm of a hot new restaurant at the time, Louro, to provide the food.

The event was eye-opening for me. We called it "Drink Like a Fish," and *The New Yorker* covered it in an article called "Seaweed on the Rocks." It was the first time I saw what creative, skilled hands could do with my products. The drinks were like nothing I could have imagined. A Thimble Island oyster shooter, made with my shellfish, included mezcal and kelp. The Green Blood Maria was made with kelp, tequila, tomato water, celery, habañero pepper, and lime. The food was completely out of the box, too. Chef Santos whipped blanched kelp into rich froths of butter and smeared it on crostini. He took the thick stems of the kelp, called stipes, and deep-fried them as tempura sticks with a bright, acidic dipping sauce. (He later whispered the secret ingredient to me: powdered ranch dressing from a big-box grocery store. No wonder I liked it so much!) He also came up with the original recipe for kelp fra diavolo, a spicy pasta dish with thin noodles of kelp standing in for the usual wheat-based spaghetti. It was a huge hit, and would go on to play a pivotal role in later events for our farm. (Check out Dave's recipe at the end of the book if you want to give it a try.)

A second major shift in my thinking came from a mysterious figure working out of the Bronx. I was on the boat one morning when I got a text from an unknown number: "How much kelp do you have?"

I looked out at the tens of thousands of pounds waving beneath the water. "A shitload," I wrote back.

"Send it to me and I can turn it into the new kale."

The guy behind the curtain was Ian Purkayastha, aka New York's "Truffle Boy." He didn't deal in seafood or seaweed but, rather, in rare Hungarian truffles, exotic wildflowers, wild-foraged herbs. He was famous for having the cell phone number of every celebrity chef in Manhattan. With nothing to lose, I shipped him off a hundred pounds, not expecting much. To my surprise, within forty-eight hours, my kelp was on the esteemed menu at Le Bernardin. What was Ian's marketing secret? He didn't see my kelp as seaweed, he saw it as sea greens, just like any other green—kale, arugula, Swiss chard—that was obscure a decade ago. To him, my kelp was just locally sourced kale of the sea. True to his word, he got sea greens into the hands of many of the top chefs around New York.

This was around the time when I stopped thinking of kelp as a seafood. I had been trying to find seafood chefs who could turn sea greens into the next big thing, but I realized that what I really needed was someone who knew how to do sexy things with vegetables. I began searching for new blood, and Brooks Headley's name kept popping up. A former drummer for the Young Pioneers, a punk rock band in the mid-1990s, he was running a little coffin-sized take-out restaurant on the Lower East Side called Superiority Burger. He rose to culinary fame making pastries at Del Posto, and was known for packing vegetables like celery and eggplant into his sweet delights. Now he had transitioned to meatless burgers. At one time, these were seen as edible only to a minor slice of consumers. Now, of course, meatless burgers are mainstream, with slop shops like White Castle and Denny's serving up finger-licking beefless burgers.

Brooks was a pioneer of the vegetarian meal that was as gut-bustingly delicious as any meaty delight. His secret sauce was making vegetables unhealthy. As a glowing review in *The New York Times* reported, he "has no interest in the veggie burger as health food, which may explain why it tastes so good. This particular veggie burger isn't tasked with alleviating whatever guilt we may be carrying around. It's a squishy little sandwich that delivers some fast, intense, umami-entangled satisfaction. It just so happens to be made with vegetables."

I sent Brooks some kelp noodles, and within the first week, he made what I see as a major breakthrough in the culinary history of sea greens: barbecued kelp noodles with parsnips and bread crumbs. In a single recipe, he completely achieved the vision I'd hoped for: de-sushified and de-healthified seaweed. Brooks used the sweet heat of BBQ sauce, the crunch of bread crumbs, and the mellow richness of parsnips to save my kelp from the quick death of the seaweed salad or as a sidekick to something fishy. He actually used it as a vegetable! Brilliant. The goal was to have people take the first bite and say, "Wow, this doesn't taste like kelp. . . ." Within a week, the kelp dish was selling out nightly. Price was good, too—six dollars a plate. Finally, seaweed for the workingman. (A version of Brooks's recipe, using kelp noodles and carrots, is in the recipe section at the end.)

In the hump and haul of hard living, regular folks deserve

delicious meals, wrapped in pleasure. Just because we're growing climate cuisine doesn't mean it shouldn't be mouthwater-
ing. Heed the words of food critics like Trip, a Rhode Island
fisherman recently turned ocean farmer: Let's make kelp the
Doritos of the future!

In 2015, I got a surprising email: *The New Yorker* wanted to
feature our farm in a full-length spread about seaweed. *The
New Yorker* had a special place in my heart, the one publication
that my Brooklyn-born mother had had delivered all the way
to the outports of Newfoundland. She read it cover to cover
every single week. It had been a ritual of mine to call her every
morning for the past two decades, and thousands of those conversations included her updates about what she'd been reading,
more often than not in *The New Yorker*. She'd even had the little
article about our kelp cocktail event framed and hung in her
home. I agreed to the new article immediately, heartbroken that
my mom wouldn't be around to read it.

The writer Dana Goodyear would be following me for a
week. (Tamanna, of course, upon finding this out, immediately
ran to the bookshelf and whipped out her copy of Goodyear's
book for me to read: *Anything That Moves: Renegade Chefs, Fearless
Eaters, and the Making of a New American Food Culture*.) We visited New York restaurants together, went out on the farm, visited with seaweed scientists. For me, the highlight was lunch at
our home, prepared by Tamanna. It would be the first of many
times we'd put Chef Dave Santos's recipes to good use: kelp
fra diavolo and kelp compound butter. We also served mussels
from the farm, which I took Dana out with me on the boat to
collect, and sea salt we had begun harvesting from the farm and
blending with delicate dried sea lettuce. She wrote about the
meal evocatively, describing the butter as looking like "the ter-
razzo floor in an old bank." In the tradition of my mom, who

had dutifully collected every single article ever written about my farm, Tamanna later had the article framed and hung it next to her desk.

A nerve-racking test for seaweed came when we partnered with the Yale Sustainable Food Program (YSFP) to host some of the top chefs in the world on our farm. The YSFP had hosted my shellfish CSF for years on their campus farm and had used my kelp as a super-fertilizer for their crops. They had launched a summer leadership summit with René Redzepi, the lauded Danish chef whose Copenhagen restaurant Noma was regularly ranked as the top restaurant in the world. René recruited a daunting list of global talent for the summit: Alex Atala, April Bloomfield, David Chang, Peter Meehan, Jessica Koslow, Kylie Kwong, Olivier Roellinger, Rosio Sanchez, and Michel Troisgros. As part of the conference, they'd spend half a day out on our farm.

The plan was to do a farm tour and then have lunch on Outer Island, the National Fish and Wildlife preserve that lies a short run south of the farm. My crew had no idea who these celebrity chefs were, so the tour was relaxed and fun. You pull lines with three thousand pounds of kelp out of the water, fifteen-foot shimmering walls of chocolate brown plants, and it'll blow anyone away.

I pulled off a few strands and dunked the dark brown blades into a thermos of steaming water, my favorite tour trick. As soon as they hit the hot water, they flash from deep opaque brown to a bright translucent green, transformed from seaweed into delicate sea greens. After a quick blanch, they're a perfect *al dente* texture, sweet with just a hint of ocean salt. René delicately nibbled a blade; Alex chomped and chewed like he was gnawing wild boar. René said, "This is a new taste, not like any kelp I've had before." The best praise you could give an ocean farmer.

They were tasting merroir, the specific flavor of sugar kelp grown in our waters. Of course, these pros were versed in the subtleties of taste. They knew better than me that the same principle of terroir—the hundreds of soil, nutrient, and weather variations that create regional flavors in wines, tomatoes, fruits—was at work in the ocean. The depths, salinity levels, temperatures all affect taste. My native sugar kelp grew in the famed freshwater-fed "sweetwaters" of the Thimble Islands, at the southernmost edge of its natural range. The effect? Thinner, more delicate, milder-tasting blades than its northern cousins, which are briny, dense, and bold.

And this is where the promise of sea greens begins to emerge. As chefs source from more farms in more regions, the complexity of the cuisine will evolve. Sam Sifton, food editor at *The New York Times,* has long been skeptical of all the hype around seaweeds, but when he began using it in combination with wintertime root vegetables, he found it "shockingly delicious" and told me he "had seen the light." A Japanese chef I met pulled from his pocket the tiniest flask I had ever seen—less than three inches high. He dabbed a dribble on my finger and nudged my hand toward my mouth. Even to my dull palate, it was exquisite. It was stunningly complex, with the depth of roasted mushrooms and the floral lightness of edible blooms. He had aged kelp broth in cedar casks for nine years.

After the tour was the island lunch of kelp fra diavolo. I wanted to prove that kelp could be an everyday home-cooked meal for the masses, and insisted on serving something that felt familiar, not rarefied. I immediately wanted Tamanna to prepare the meal, but she resisted, as I knew she would, never confident about her prowess in the kitchen, and, among all of us, she was the only one who had any idea who all these people were. But when I said that I would just do it myself, she looked at me in horror and kicked me out of the kitchen. She spent all eve-ning stirring and stirring that big pot of sauce, carefully adjust-ing seasonings. And, as I expected, she kicked ass. René came back for seconds, and David Chang had thirds. As the meal was winding down, Chang came over to me and said, "Fuck it, let's do this."

By that point, we were confident the chefs wanted the prod-uct, but the ultimate test in my eyes was whether regular folks would want it, too. We'd find out at the Quinnipiac River Fes-tival, in my own neighborhood of Fair Haven in Connecticut, where we planned to serve samples of kelp noodles. Fair Haven is a rare community for the East Coast: less than two square miles, bounded by two rivers, with two of the largest highways in New England looming over it, Routes 95 and 91. I feel as if Fair Haven failed to get the memo that we're supposed to have a "race problem" here in America. It's 50 percent Latino, 25 percent black, and 25 percent white, mostly a mix of poor and working-class families, and everyone seems to live side by side in a totally mixed-up, vibrant community. My Angelena wife says it's the one place around here that makes her feel at home. It's in the same town as Yale, but feels a million miles away.

I expected a tough audience for something as exotic as kelp. It was a bright, sunny morning, and beer was already being poured. There were tables set up for diabetes prevention, the food bank, a mobile library. A raffle for hundred-dollar gift cards at C-Town, the local supermarket. My table had some Thimble Island Ocean Farm shirts, buoys, and scallop nets for décor, and little Dixie cups piled up with our trusty kelp fra diavolo noodles, with big chunks of enticing shrimp to draw people in.

"What's that?" asked a guy who said he was a roofer.

"Kelp noodles."

"What? What did you say?" he nearly shouted; he had that singular edginess of roofers I know, always a touch crazy.

"Kelp noodles in tomato sauce."

"You eat that shit?"

"Yeah," I said. "People in New York like it. Yalies, too."

"Those fucking people will eat anything."

"Don't be a coward—try it."

He scoffs, but now his roofer's honor's on the line. So he takes a bite. "That ain't too fucking bad. Ain't too fucking bad at all." He ate two more. Nailed it again.

I figure half the folks that stopped and tried the kelp had similar reactions; the other half politely said it was "interesting." Those are good numbers. It means kelp is polling at 50 percent, and that's a hell of a lot better than the president or Congress. By noon, the roofer, wobbling drunk, had come back for thirds. He had to fight off a long-haired hippie who was shoveling six, seven cups at a time into his mouth.

So now we knew: celebrity chefs, journalists, foodies, roofers. Kelp really could have mass appeal.

FALLING IN LOVE WITH SEA GREENS

The pinnacle of my transformation from seaweed-hating ocean farmer to true believer happened in Copenhagen. In 2017, I was invited to speak at a MAD gathering, a nonprofit think tank launched by the chef René Redzepi that is dedicated to transforming the global food system. Tamanna came as well, and we spent the most romantic week of our lives, eating and drinking our way through the snowy cobblestone streets of Copenhagen. René was kind enough to save us a table at his restaurant, Noma, typically an impossible reservation to secure.

It was truly magical. From the often-written-about full-staff greeting at the door of the restaurant to the exquisitely prepared dishes, it was unlike any other dining experience. We were both nervous, out of our element, but even my McDonald's-trained palate could appreciate the brilliance behind each plate. René and his cooks had done it, everything that I had been promising about the future of food: Moving sea greens and bivalves to the center of the plate and fish to the edges. Making sea vegetables uniquely delicious in their own right. Finding the connective tissue between soil and sea.

The level of creativity was unreal, and it tasted good, too. Charred greens with a scallop paste. Sea urchin and cabbage. Langoustine, onion, and lavender. Moss cooked in white chocolate. Radish pie with crust made with roasted kelp. Twenty courses of this, with sea vegetables effortlessly playing a starring role.

I later wrote to thank Noma's staff, and the head server sent back the following descriptions of two of the dishes we'd been served that day.

Cooked oyster and broccoli stems dish: Along with the broccoli stems, blackcurrant buds, thyme and black currant leaf oil on top of the oyster is also some kelp salt. This kelp salt is the process of making a kelp broth, extracting all of the natural salts out of the seaweed, but then dehydrating this broth to be left just with the pure crystals of the kelp salt. Very intense, and umami rich.

Roasted bone marrow / Maitake mushrooms: The grilled mushrooms are brushed with a glaze of the reduced kombu, dried cep, trumpet and morel mushrooms, lingonberries and lactic cep water. This is brushed over the mushrooms as they are grilled outside over charcoal.

Purple brussel sprout / cabbage leaves that the bone marrow or mushroom are eaten in (which act as the wrap) are seasoned with many things including horseradish juice, fennel or parsley flowers, unripe walnuts but also cubes of kelp—this kelp is cooked for a couple of days in a broth with dried mushroom and berries until very tender and then sliced into small cubes. Vegetable reduction, or the brown sauce, is made by a reduction of the roasted kelp dashi, fennel juice and a paste of black garlic, mattock mushroom and an elderflower pea-so.

This is just a glimpse of the level of care and attention taken with every ingredient served. The meal was one we will never forget, and unlike anything we are likely to experience again. An inspired and beautiful example of what climate cuisine can look like.

When I read Goodyear's book, it was a crazy ride through bugs, dirt soups, and habañero cotton candy. What struck me, though, was her analysis of why and how food tastes change. One influence on taste is environmental. According to Daniel Pauly, a fisheries ecologist, "This idea of 'liking things' is actually a reflection of the pressure on the environment. You like what you can get. I grew up in Switzerland eating lots of horse meat because it was cheaper, and so I liked horse meat."

Over time, seaweed and shellfish are going to become cheap, too. Since our crops are zero-input crops—requiring no fertilizer, feed, land, or freshwater—the emerging pressures of the new climate economy will make them the most affordable food on the planet. Producing zero-input food will allow better prices in the grocery store, driving consumers watching their wallets to the sea greens aisle. What started out as haute cuisine will, and must, become a new dinner favorite. The Gorton's Fish

Sticks of the future. This has always been the way of things: food prices have a powerful effect on overcoming taste barriers. The question we face now is whether we can introduce sea greens as a new food product that is both delicious and beautiful, and I think the answer is yes.

In her book, Goodyear uses America's adoption of sushi to show that changing tastes are driven by more than just resource limits. Initially, when sushi was introduced to the United States in the 1960s, the assumption was that the American public would never accept it, mainly because it is raw and handled without gloves. According to Goodyear, the convergence of three factors changed their minds: "the food pyramid, which emphasized fish; the rise of the Japanese car; and *Shōgun*, the best-selling novel by James Clavell."

So—we're putting our eggs in the sea vegetables basket because they hit the holy trinity needed for the future of food: delicious, nutritious, and restorative. How often will we be granted the opportunity to explore an unexamined ecosystem right outside our back door? If we can just convince moms of the future to scold their kids to "eat their ocean vegetables," there is hope of both saving our seas and feeding ourselves. Maybe all we need is a tweak to the food pyramid, a page-turner, and a self-driving car called the Kelpie. Sounds doable to me.

CHAPTER 15

Sea Greens for the Masses

Through the process of learning about kelp and the experiments with chefs and local events, I had come to soften on sea greens. Nobody was more surprised than me. I'd sworn I'd never eat the stuff; I regularly irked other seaweed companies by proclaiming publicly how disgusting it was. But, with the help of my wife (and the world's top chefs), I was becoming a convert.

But I knew that a few upscale restaurants in New York weren't going to create a market that could sustain what I hoped would be a new economy for out-of-work fishermen. And it certainly wasn't going to make a true environmental impact. My farm was doing great—I had a dedicated following for my monthly CSF shares, and high-end restaurants were buying all the shellfish and seaweed I could turn out. Now I had to find a way to bring this vision up to scale.

KELP IS THE NEW KALE

"Kelp is the new kale" has long been my hopeful tagline—but how did kale become kale in the first place? I'm a poor litmus test for what is popular in the health-food world, but I have no recollection of ever hearing about kale a decade ago. Even my veggie-loving wife doesn't remember that kale wasn't so popular until relatively recently. How did it happen?

As I looked into this question, I learned that kale's ascendency was no accident. It came with a cloak-and-dagger playbook developed by a PR hired gun named Oberon Sinclair. Before the "Kale Queen" showed up, the main buyer for kale was Pizza Hut, which used the leaves as a garnish on their salad bar. Sinclair changed all of that. She claims to have been hired to popularize kale by the American Kale Association, which turned out to be a fake organization that she had started. No one knows where the dark money came from, but the campaign was multifaceted and hugely successful. Between 2011 and 2014, American search queries for kale quadrupled, sales skyrocketed, and producers of kale chips became overnight millionaires.

Rumor has it that kale was "discovered" in 2011, during an Underground Culinary Tour in New York set up by Sinclair. It was a secret, invite-only event, with guests required to be either celebrity chefs or executives at major restaurant companies. They were taken in a stretch limousine to visit fifteen restaurants in an epic twenty-five-hour tour. Two of the guests reportedly fainted from exhaustion. The purpose was to expose key influencers in the food industry to cutting-edge products and food trends.

Jeffrey Frederick, the vice president of food and beverage at Caesars Entertainment, which is the largest gaming corporation in the world, was one of the tour guests that day, and kale salad was served. Frederick, the rainmaker behind Gordon Ramsay and other celebrity restaurateurs in Las Vegas, returned home and used his influence to push kale to his chefs. They resisted the texture as too fibrous, the taste as too bitter for the American palate. It was just too weird a food.

Frederick told them to begin using it as an ingredient anyway, and within a few months kale began popping up on menus all over Las Vegas, and soon after the rest of the country. Others quickly joined the movement. The psychiatrist Drew Ramsey, a professor at Columbia University's medical school, cofounded National Kale Day and wrote the book *Fifty Shades of Kale*. His nonprofit helped serve kale to twenty-five hundred schools and eventually strong-armed the U.S. Department of Defense to dish out kale at twenty-four hundred commissaries on National Kale Day.

Fast-forward a few years and kale was surfacing as a hot baby name, farm level production had risen 60 percent, and there was a 400 percent increase in its appearance on restaurant menus. McDonald's even began testing kale breakfast bowls! So a vegetable that had been considered feedstock for rabbits was suddenly treated as a thrilling new ingredient, and now, a decade later, it's competitive with spinach in the grocery aisle.

So, with a little instruction in how to create a market for something new out of nothing, I got to work. I printed up some T-shirts declaring "Kelp Is the New Kale" and had Chef Dave Santos—who had become a good friend and adviser by this point—simplify the recipe for kelp fra diavolo that we had used for events in the past. He nailed it with a simple and accessible recipe: kelp noodles, tomato sauce, cayenne pepper, and shrimp. Dave's dish became my gateway drug.

We began to shift our focus from high-end restaurants to high-end institutional buyers. Seaweed grows like, well, weeds, and if we wanted to move all the kelp future ocean farmers would be growing, we needed major purchasers. I started by reaching out to the food team at Google, who had previously invited me to speak at their yearly conference.

Tech companies have lately become infamous for the array of perks they offer on-site in order to attract and keep top talent, from dry cleaning to car washes to pet grooming to child care. Facebook even hired Disney consultants to design its sprawling California campus, which includes nine restaurants, a bank, a barber, a music studio, and a video arcade—all free for employees. But Google was the first to introduce the idea of the high-end employee cafeteria, offering free gourmet breakfast, lunch, and dinner daily to its thousands of employees, including wine and beer for Friday happy hours.

Google not only set the standard for the quality of the food, but went a step beyond, actively seeking to improve the health of their employees with subtle nudges toward healthier choices. Employing a cadre of food psychologists, nutritionists, and chefs, they explored how packaging color affects food choices, offered smaller plate sizes (which has been proved to help people eat less), and put healthy snacks like fruit up front and sugary snacks farther from reach, in opaque containers. Soda is available, but it's at the bottom of the fridge case, behind frosted glass, with water front and center. When Google does something, others pay attention; many of these little health experiments that proved successful for Google employees have now become commonplace in corporate cafeterias beyond the tech sector. We knew that, if we could get a trendsetter like Google onboard, it would be a major win for the future sea veggie farmers of America.

Google's culinary team traveled to the Thimble Islands for a tour of the farm. They arrived in a huge unmarked black bus, with tinted windows and music playing. All the people who poured out the doors looked as if they had just arrived from a wellness retreat, gleaming with health. They brought with them buyers from the Compass Group, the largest contract food-service company in the world. Compass provides food service for thousands of high-volume corporate cafés, hospitals, schools, arenas, museums, and more. Their clients in New York alone include the Metropolitan Museum of Art, the United Nations, the World Bank, the Google headquarters, Morgan Stanley Global Headquarters, and the New York Aquarium. These were high stakes.

We had debated among ourselves what to serve these food professionals. A smorgasbord of tapas-style small plates? Tempura stems, pickled greens, blended salts? In the end, we decided that, if we truly wanted to make sea greens an everyday food for the masses, it was important to showcase them at their most accessible. We stuck with our tried and true formula of kelp noodles, served in our usual low-key style—on a folding table on the docks, heated up on the Coleman stove I had used when I proposed to my wife out on a barge.

The group unanimously loved the dish. As they mingled, following the talk I gave, one after another came up to me, commenting in shocked voices, "This is actually *really* good," and "I could see eating this every day." They were surprised to see that sea greens could be a versatile, neutral ingredient, and that they might be used as a main dish, not just a side. Mike Wurster, Google's brilliant culinary director for the East Coast, has been our biggest champion and advocate since that day, serving tasty burgers made half of kelp and half of beef, and introducing sea greens into cafeteria menus. (Fun fact: McDonald's was actually the first to pioneer the seaweed-based burger—the McLean Deluxe, which was 91 percent fat-free. Amazingly, it lasted on the menu for five years, and was the official sandwich of the National Basketball Association!)

As more and more kelp hits the docks, the goal is to take sourcing to the next level by engaging Compass and other major buyers who can bring sea greens to the broader public en masse. Institutional buyers like this have a major role to play, because they determine what people eat in the places where they work, play, learn, and do business. They can help normalize sea greens as just another vegetable option on the menu.

The support of public entities is also important for a growing industry, and a few years ago we got a great nod of support from the Obama administration. On June 16, 2014, at the "Our Ocean" Conference, the White House served my American-farmed seaweed for the first time in the history of our country. Barton Seaver, global culinary ambassador for the United States and master chef, prepared the meal and presented it to Secretary of State John Kerry and a hundred global leaders: "I am serving you chilled kelp coleslaw tonight. This is the food of the future. I guarantee in the years to come, we will all be eating locally grown sea vegetables."

Standing before the political class, Barton was pointing to a new seafood plate, reimagined with ocean plants and bivalves at the center and wild fish pushed to the edges. I'm convinced that chefs like Seaver and the others we have worked with will help make our new ocean "climate cuisine" creative and delicious, and we can be the trendsetters to make mainstream food companies take notice. Kelp is now poised to be the ambassador for an entirely new cuisine.

SOY OF THE SEA—JUST NOT EVIL

Critical to the strategy for creating a new economy, and a "just transition" for fishermen, is focusing on more than just food. Growing numbers of ocean farms will mean that millions of pounds of kelp are harvested each year on the East Coast alone. Even if every American ate sea greens every week, it wouldn't be enough—where is all the rest going to go? Interestingly, learning about the soy industry has shown a way forward.

In the 1950s, soy faced a similar challenge. Farmers had found that soy was fast and easy to grow, and could prove to be a huge cash crop. But Americans weren't exactly chowing down on tofu. There had been numerous attempts, from as early as the 1930s, to overcome the bias against soy. Big Soy had poured millions into piloting new products. Madison Foods, a Tennessee-based soy manufacturer, launched the Soyburger in 1937, a canned soy-wheat loaf filled out with raw peanut flour. It was reportedly just as bad as it sounds. Undeterred, Loma Linda Food Co. rolled out the Gluten Burger in 1938. It was, as *The New York Times* recently recalled, "one of the many shelf-stable cans of soy-based meat mimicry that has given veggie burgers a bad rep over the years."

The agents of soy tried again in 1942 with an article by the Los Angeles–based health-and-wellness guru Mildred Lager in her newsletter, *House of Better Living*, entitled "Soy Bean Recipes: 150 Ways to Use Soy Beans as Meat, Milk, Cheese, & Bread." Unsurprisingly, *Soybean Digest* gave it a glowing review. Nobody else paid much attention.

So Big Soy shifted their strategy. They held an industry-wide meeting in the 1950s—in what I imagine to be a shadowy and smoke-filled room—and decided that marketing soy wasn't working, so it was time to go underground and sneak soy into everything. And they meant *everything*. If you fast-forward a few decades, you can find soy in margarine, particleboard, vitamins, cereal, infant formula, frozen desserts, bedding, lipstick, pet food, soups and stews, mayonnaise, peanut butter, seafood, chewing gum, candy, nutritional supplements—the list is endless. A recent advisory for those with allergies warned of seventy-eight different names that soy might be hiding behind on your food labels. Soy is now the largest source of protein for the world's farm animals. It is being used for fuels, for construction materials, as ink. Mission accomplished, Big Soy.

It's been nearly seventy years since that fateful meeting, and they're still at it. In 2014, I was asked to attend a gathering pulled together by the Clinton Global Initiative. I have no special love for the Clintons, but I figured I needed to be in the room so I could continue to learn from all the folks innovating in the aquaculture space.

Instead of farmers, though, the room was filled with big, hulking Midwestern soy-industry reps. I didn't get it: Why were so many land-based soy producers at an aquaculture conference? So I went up to a herd of them grazing on the pastries: "What are you all doing here?" The answer: "Fish feed." Ahhhh, it all clicked into place. Aquaculture is the fastest-growing agricultural sector in the world, and they want in on the game. Keep with the master plan—put it in *everything*!

The problem with all this is, soy sucks. Soy production is responsible for the clear-cutting of millions of acres of forest, savanna, and grasslands. Not only does this destroy habitat for innumerable species, it also has a devastating impact on local communities who depended on these lands for their survival. It has a major impact on climate change. It's a water hog and a root cause of much of the soil erosion plaguing global farmlands. It's grown in massive monocultures whose maintenance requires the heavy use of chemical fertilizers, herbicides, and pesticides, toxic substances that contaminate soil and waterways, making a major impact on wildlife and human health.

In addition, 80 percent of soybeans grown in the United States are genetically modified to tolerate glyphosate (better known under the brand name Roundup). When the amount of residue left from these chemicals was found to far exceed the allowable limit by the FDA, instead of cracking down, the FDA did exactly the opposite: it tripled the allowable limit. This increased limit remains in place despite evidence over the last several years that glyphosate is a likely carcinogen. Recently leaked FDA emails showed that glyphosate is now detectable in many everyday food products. In 2017, *Time* magazine reported that glyphosate is increasingly being found in people's blood tests.

In short, soy is destroying the planet and our health.

Learning about all of this got me thinking. We're creating an industry based on a food product, seaweed, that is exactly the opposite of these large, destructive mono-crops, yet also has the potential to turn out massive quantities of product. Kelp replenishes ecosystems, rather than depleting them. It helps mitigate climate change by sequestering five times more carbon than land-based forests, earning the moniker "Sequoia of the Sea." It requires zero inputs—no feed, no freshwater, no fertilizers. It creates rich new habitat for all kinds of species, as all the fishermen who enjoy angling on my farm can tell you. So why not take a page from Big Soy's playbook and find new and innovative ways to use kelp in different industries? Instead of going underground to hide a problematic product, we can do just the opposite. We can shine a spotlight on all the varied uses of a plant that just may help us save the planet.

BEYOND FOOD

Buoyed by this new inspiration, I now had a two-pronged strategy for the fledgling sea-greens industry. We were well on the way to increasing the culinary uses of kelp, but we could also make kelp the soy of the sea by weaving seaweeds into as many other industries as possible. For hundreds of years, humans have found nonculinary uses for kelp. In the eighth century, Islamic scientists extracted the compound algin from a brown seaweed, and used it to fireproof boats; today, firemen's clothing is still treated with alginate for fireproofing. Seaweeds were used as gunpowder in the Revolutionary War and chemical weapons in World War I (let's skip those uses this time around).

In my local post office, there is a Depression-era mural painted by an artist commissioned by Roosevelt's Works Progress Administration, depicting farmers with wheelbarrows scouring the shore for seaweed to fertilize their fields. Long before that, the Scots were pouring vats of porridge into the ocean as an offering to the ocean gods in hopes of a good seaweed harvest, which they used to enrich their oat fields.

With kelp, we can revive the long tradition of using seaweeds to break down the walls that separate our land-based and ocean-based food systems. We know now that even the best land-based farms pollute, sending nitrogen into our waterways; in fact, 85 percent of fertilizers eventually end up in our oceans, streams, and lakes. So, like the Scots, we can use our kelp to capture that nitrogen, turn it into fertilizer, and send it back to organic farmers to grow their vegetables. When the nitrogen then runs back into places like Long Island Sound, we capture it again. The idea is to build a bridge between land and sea in order to close the loop between our food systems. Too often our thinking stops at the water's edge.

From the ocean farmer's perspective, fertilizer creates a market for "waste." One of the challenges of kelp is that, when it's harvested for food markets, it can be tricky to get the timing right. If we keep our kelp in the water too long, snails attack, leaving holes; bryozoans cover the blades in white blotches; and competing seaweeds use the kelp as a surface to grow on. When

I first started farming, I'd trash this biofouled kelp, because chefs didn't want kelp with tons of other shit growing on it. But for fertilizer markets, that doesn't matter. In fact, all that biofouling represented additional nutrients that are loved by land-based farmers. This meant that, the longer we left our kelp in the water, the more potent the fertilizer as a product.

Here in New England, we've already built partnerships with local farms that use our kelp as an organic fertilizer. The Yale University farm reported increases in yields after a season utilizing our kelp, and Gentle Giant Farm has begun creating kelp compost. Seaweed fertilizers are already well known to farmers and gardeners, and recently a major organic fertilizer company reached out to source five hundred metric tons of farmed kelp. Fertilizer could prove to be a major use for our seaweeds, and it's exciting to consider the prospect that each coastal state might grow its own local fertilizers for parks, schoolyards, and community gardens.

Growing vegetables isn't the only use of seaweed in agriculture. It can also be an important source of animal feed. The archaeozoologist Ingrid Mainland has confirmed that the use of seaweed as a fodder for sheep and cattle in the Orkneys, which still continues today, dates to the Neolithic Period, roughly five thousand years ago. The result is umami-rich meat akin to the salt-marsh beef famed in France. Not only can feeding animals kelp produce unique and tasty meats; it turns out that it can also be a weapon for fighting climate change. Cows produce an alarming amount of methane when they expel gas. Methane is considered to be about thirty times worse than carbon dioxide in contributing to global warming. Livestock produce more greenhouse gases than global airplane and car emissions combined, and 65 percent of these gases come from cows.

In 2016, exciting research came out of Australia's James Cook University showing that the addition of a small amount (less than 2 percent of the total feed) of a macroalgae called *Asparagopsis taxiformis* reduced methane production by 99 percent. More recently, in 2018, Joan Salwen began taking this research to the next level by building a network of leaders from academia, the dairy and beef industries, and ocean farming to put the research into action. She grew up in rural Iowa, "in the long shadow of [my] family's 100-acre farm," as she says. "[My] Grandpa King produced pasture-raised cattle, soybeans, and hardworking off-spring." Now Joan is reimagining the American family farm for the era of climate change by exploring the full potential of *Asparagopsis* as a methane-reducing livestock supplement.

Her nonprofit funded researchers at the University of California in Davis, a renowned agricultural university, to test the Australian research findings in live cattle. The preliminary results showed a nearly 60 percent reduction in methane production, far beyond the 30 percent that was initially hoped for. This is astonishing. It's significant that Joan is working to make sure farmers are central. "Farmers and ranchers are part of the design process," Joan says, "And their stories and aspirations are important."

Joan and I share a larger vision as well. As Joan puts it, "A cooler Earth is one in which less greenhouse gas is emitted, but also one in which economic opportunity and justice is fostered." Joan is committed to making sure all of the knowledge gathered is open-source and available to farmers. A recent *National Geographic* article noted that, if this proves to be a viable method of reducing greenhouse gases, there would need to be a massive increase in seaweed farming to meet demand. Sounds like we're going to need a lot more ocean farms!

Biofuel is another area we've just begun to explore. Finding a clean replacement for existing biofuels is becoming increasingly urgent. A report commissioned by the European Union found that biofuels from soybeans can create up to four times more climate-warming emissions than equivalent fossil fuels. Seaweed and other algae are increasingly looking like a viable substitute. About 50 percent of seaweed's weight is oil, which can be used to make biodiesel for cars, trucks, and airplanes. Scientists at the University of Indiana recently figured out how to turn seaweed into biodiesel four times faster than other biofuels can be made, and researchers at the Georgia Institute of Technology have discovered a way to use alginate extracted from kelp to ramp up the storage power of lithium-ion batteries by a factor of ten.

The DOE estimates that seaweed biofuel can yield up to fifty times more energy per acre than land crops such as corn. The world's energy needs could be met by setting aside 9 percent of the world's oceans for seaweed farming. "I guess it's the equivalent of striking oil," says Tasios Melis, a professor of microbial biology at the University of California, Berkeley. Closer to home, a high school student at the Bridgeport Regional Vocational Aquaculture School, right up the road from our farm, recently used our kelp to create a method for recharging a twelve-volt battery, winning a national science competition!

Looking at the larger picture, the potential of ocean farming to disrupt existing systems is massive. Through reimagining our oceans as productive spaces growing not only food, but fertilizer, animal feed, and fuel, we are on the cusp of something truly revolutionary. What if the ocean could offer a way to rewrite the mistakes of our past?

FARMING THE URBAN SEA

I often get asked if it's possible to farm in urban areas. The answer is yes, and we're already doing it. We call it pollution farming, purposefully siting farms not for food production but for bioremediation—what scientists call "ecosystem services"—near metropolitan areas. The goal is to leverage the power of shellfish and seaweeds to filter water and soak up carbon, nitrogen, and heavy metals, all the while rebuilding reef systems.

An example is Dr. Yarish's groundbreaking work in New York, where he set up a kelp and shellfish farm in the Bronx River to pull nitrogen, mercury, and other pollutants out of the city's waterways. Amazingly, the kelp grown in the Bronx River actually passed FDA food-safety tests. I can only imagine how hard it would be to market seaweed grown at the tip of Hunts Point, but these results show that some urban farms' crops will be food-grade. Others will be suitable as fertilizers or feeds, and the really nasty stuff can be turned into biofuel.

The promise of pollution farming has begun to emerge as a reality. In collaboration with the Port of San Diego in California, Sunken Seaweed, started by Leslie Booher and Torre Polizzi, built the nation's first commercial urban farm and seeded five varieties of seaweeds, with an eye to reviving the kelp fertilizer and animal feed processing plant that had operated on the docks during World War I.

What's especially exciting is that these farms have the potential to generate revenue by "harvesting" what are known as nitrogen and carbon offsets. These offsets enable companies to reduce their environmental impact by supporting projects that reduce or absorb carbon and other emissions in order to compensate for an equivalent emission entering the atmosphere. So, when a company purchases a carbon offset (from, for example, ocean farmers), it will lower their carbon footprint, while also creating new revenue streams for farmers. Locally, we've been pushing for an expansion of Connecticut's existing nitrogen credit trading program—currently the only one in the country—to include shellfish and seaweed farms, thereby reimbursing ocean farmers for the nitrogen they filter from Long Island Sound each year.

Rather than just patting ocean farmers on the back for their good work helping the environment, repositioning markets to reward farmers for their role in protecting the planet has the potential to be a major driver for getting more farms in the water.

Keeping Your Farm Afloat

As long as you don't mind farming when it's below freezing and blowing twenty knots and you have to break through ice floes to make it out of the harbor, maintaining the farm is pretty easy, although some crops take more work than others.

Your job as an ocean farmer is ensuring that your crops consistently get the right mix of sunlight and nutrients. This means using the right mix of floating and gear tensioning. Over the years, I've found that each crop enjoys a different level of the water column: kelp at the top, mussels and scallops in the middle, clams on the bottom, and oysters at any of the three levels.

We have two ways to control depth, and both are simple. One is adding or subtracting the number of buoys on the horizontal longlines as the seed matures and gets heavier.

The second is to tighten or loosen the longline. This is especially useful if storms are expected to roll in, because you can lower the farm below the big rollers. It's also helpful at the end of the season, as surface temperatures rise. We loosen the kelp lines, sinking the kelp down to cooler waters. This slows down growth and also slows the rotting of the kelp blades, which in turn extends the window for harvesting.

You should "walk" the lines at least once a month, checking for fraying and other wear and tear of gear. In the beginning, you should inspect all your connections weekly, to make sure everything stays in place. It is a good idea to start a log to document your inspections of the farm. Out of all the crops, oysters and scallops require the most work, because the lantern nets and oyster bags get biofouled. This means that all the critters and plants of the ocean attach to the mesh, suffocating the shellfish. In the summer, when biofouling is the worst, I rotate out my lantern nets and mesh bags so they can dry in the sun and I can easily scrape off the dead fouling. Oysters also need to be tumbled—basically, just roughing them up, in order to thicken their shells and deepen the cups. We built a simple hand-tumbler that looks like the rotating compost bins that gardeners use. I tumble my oysters three times a year.

If you're farming in cold climates, ice is always a challenge. In the last fifteen years, my farm has been destroyed two times by ice floes, including one time when the entire infrastructure of the farm was dragged half a mile south. Nightmare. Now, for especially cold years, we switch out our floats with "winter moorings," commonly used in marinas. They are basically just cone-shaped floats that keep ice from gripping the buoys and dragging the farm with it.

Of course, there's plenty more to learn, but these are the basics. Remember, pick your weather days—it can get rough on an ocean farm, and cold water immersion is a real danger.

part 5

CHAPTER 16

The Green Wave

B y 2014, I had grown a blue thumb and learned to read my waters in new ways. Through oysters, I'd gotten to know every inch of my farm's seafloor, and by growing plants vertically, I learned the patterns and seasonal variations of the full water column. This meant daily observation of how the currents moved through my fields, where the sweetest nutrients flowed in the water column, how a mix of winds and tides shifted the movement of my kelp rows.

I was now farming for taste as never before: pruning out the younger, translucent plants for chefs who wanted hints of sweetness and delicate mouth feel. Harvesting the stipes, late in the spring, after the blades have already gone to market, in order to get the thickest and longest stems for pickles. Watching the water temperature for that precise moment when the flesh of the plant is darkest, which would in turn produce the brightest green color when blanched.

My inner life was changing, too. I was fully reborn as a farmer. The hunter in me was gone, and I had lost my thirst for blood. My brain felt different, like it had been rewired. Others might not have noticed, but I was calmer, with fewer thoughts piling up in my mind. My need for high-octane adventure waned. My shift happened slowly, born out of watching kelp grow. I can't put my thumb on it, but there was a mystery to growing plants, a pleasure apart from fishing and farming shellfish. Maybe it was kelp's iridescent colors, maybe the wonder of hoisting fifteen-foot walls of plants out of the water. I imagine this is the sweet quiet that inspires poet-farmers like Wendell Berry.

Seaweed had changed me.

I was loving my rhythmic weeks on the farm and the steep learning curve. Ocean farming required a fascinating mix of engineering, science, policy, and business. I decided it was time to scale production by tripling the size of my farm. Rather than ask the banks for money, I set up an online fund-raising campaign on Kickstarter. Framing the message for the campaign wasn't easy. As a former fisherman, I knew how to spin a good tale, but I struggled to figure out how to tell my story to a wider audience. I could do a food-centered campaign, with my tagline to "make kelp the new kale"—but that wasn't what moved me. It was climate change that was on my mind, and on everyone else's, especially in the wake of Irene and Sandy. But that was a tale of doom and gloom. I had a hunch that a story of hope, of someone rising from the troubled seas, could break through the malaise. So I titled my page "3-D Ocean Farming: Saving Our Seas," followed by the tagline "Overfishing, climate change, acidification—our oceans are in peril. But there's hope. Let's start a Blue Revolution!"

A bit cheesy, but it worked. I raised nearly forty thousand dollars, blowing past my fund-raising goal. The haul of money was great, but it wasn't the most valuable result. As the thirty-day campaign picked up steam, hundreds of people began to reach out, very few of whom were ocean or seafood people. Soon it was a flood of thousands. This was in the early days of Kickstarter, when it was still novel to crowdfund an idea; they put it on the front page, and somehow word spread like wildfire. Architects and landscape designers viewed what I was doing as underwater design; Purina Pet Food asked to meet to explore the nutritional potential of sea plants for pets; Unilever wanted to talk about sourcing for cosmetics. The founder of the Paleo Diet donated five hundred dollars, and a set designer for a Marvel Comics *Spider-Man* sequel wanted some kelp so the supervillain could rise from the depths wreathed in seaweed. An anal-sex lube start-up even wanted to beta-test some locally sourced seaweed.

Soon stories began running on NPR and the BBC, in *Fast Company*, all about the revolutionary potential of 3D ocean farming. It was crazy, and a total mystery. For years, I had just been treated as a curiosity or, at best, experimental—my shell-fish co-op called me "The Petri Dish." My goal was simply to grow as many crops as possible on twenty acres by mining the past for every positive tributary of aquaculture that I could find. It was a modest endeavor to keep me on the water, but now I was riding a wave propelled by some unknown force.

Then the crazies began to show up—and it got weird. Thomas Harttung, a Danish farming and food pioneer who became one of my most valuable mentors, explained this was inevitable. "Whenever you come up with a new idea, it attracts crazy people. This is because, more than anyone else, crazy people are open to new ideas. They are not risk-averse. The problem is, they are crazy."

They came in droves. There was an Austrian guy who claimed

to have convinced the Caribbean tourist industry to support building 3D farms along their coasts in order to attract tourists and protect their beaches from storm surges. He talked me out of two thousand dollars of my fund-raised cash to set up a farm, and then vanished—my version of being suckered by a Nigerian email scam. There was the Japanese "entrepreneur" who wanted me to support his geo-engineering project of dumping billions of pounds of iron into the ocean to increase carbon capture. One woman wrote in to the website with an urgent subject line: "Protect the Sharks!" She pleaded:

> Can we make farms so that sharks, and fish, won't get lost in them? Unlike a cornfield, there would be clear pathways out, sections that were much wider, or circular, every ten rows or something? It seems that sharks could get lost.

Right. Because sharks have never encountered seaweed before. Another wrote asking if she could use our farm as a birthing center so babies could be born in the ocean, wrapped in kelp, with dolphins as midwives. God help me.

But there were also regular folks, mixed in with the scammers and kooks, who legitimately wanted to become famers. Out of the mist came lobstermen, gillnetters, college kids, burnt-out cops, disillusioned lawyers, stay-at-home moms. I spent countless hours answering emails; I felt an obligation to all those who had reached out to me, wanted them to know they were heard.

There were calls from young land-based farmers who couldn't afford farmland, veterans just home from the Iraq War, First Nations people with deep connections to land and sea. Many were heartbreaking tales. One West Coast farmer turned electrician wrote: "I've lost everything. My girlfriend of 12 years passed away 8 months ago and my family farm I live on, and the house

I built with my two hands is being sold. I love the ocean and rely on her beautiful bounty for sustenance. All I got left is my boat, my family, and my black lab. Please help me."

And one handwritten letter from a prisoner on his own journey of redemption still hangs on my wall:

Dear Mr. Smith:

I have so much to say in this letter, but in the interest of your valuable time, I'll do my best to get my point across in the least amount of words as possible.

I am a native South Eastern Floridian, but currently reside in the state of Massachusetts. I grew up in Florida enjoying the outdoors, water, and in particular, the ocean and have long dreamt of the possibility of making an honest living in conjunction with it. In 2003 I bought the cheapest ticket I could get (Thanksgiving night) and flew to Seattle WA, with only my suitcase, 1200 bucks, a reservation for a hostel within a bus ride of "Fisherman's Terminal" and a desire to get my foot in the door of the fishing industry. For the purpose previously stated I will forgo the details of my experience, suffice to say that with earned wisdom I have discovered I am a conservationist at heart and maintain the utmost reverence for the ocean, nature and fellow human.

Since first learning of ocean farming and its modern techniques it has peaked my interest as a prospective honest and noble business model that has the potential to be both profitable and beneficial to the environment. I have eagerly sought out and absorbed every bit of information that I could find on this subject, but am experiencing difficulties gathering additional information as I am dependent on others research and send it to me. I am currently enrolled and Maintaining a 3.26 G.P.A. in a community college course in "Small Business Man" that I have received a Federal grant for. I am doing my best to plan my re-entry to society where I will be expected to seamlessly assimilate without permanent residence or money, save the gracious support of two childhood friends, brothers actually, who have joint ownership of a successful painting business, who have pledged to continue to look out for me in terms of a job, temporary housing, and possibly a small bit of seed money for my own "S.B."

I am writing this letter out of exasperation of my currently helpless state and with the understanding that in order to get any assistance I must reach out for it. I have learned over my 39 years of life to temper my expectations, but if you were able to find the time in your undoubtedly busy schedule to read and consider my correspondence, then I would consider this endeavor a success. . . . Are there any available mentorship programs or assistance I can apply for?

To provide a bit of context, again, in the least amount of words possible. I am currently and have been for nearly a decade incarcerated for a crime that not only devastated countless lives, but ended prematurely the lives of my two friends, and my fiancé and mother of our one month old at the time son Jason Jr. All passengers in a vehicle I was driving while intoxicated on our way home from watching fireworks on the 4th of July 2009. I have no defense or excuse. I knew better.

I was indifferent to the 10 to 12 year sentence I received as a result as I had already began a life sentence of this almost unbearable burden. But I have determined through constant self-assessment and maturation in conjunction with the unconditional love of my family and friends, that knew my true character, which I can only aspire to deserve one day, that I still owe it to my son and them to try and forge onward, in order to pay their love forward.

Thank you for reading this . . .

I didn't know it then, but this was a taste of what was to come. There was just something about restorative ocean farming that was magnetic to the masses. My little Kickstarter campaign was accidentally filling holes in people's hungry hearts.

But why? I suspect it has to do with a widespread hunger for agency. For so many of us, the problems of the world feel insurmountable and the solutions unattainable. Only the Googles and Amazons of the world have the heft to bankroll our solutions to the climate crisis or food insecurity. But the secret message behind my tagline of "20 acres, a boat, and $20,000" is meant to make a solution concrete and attainable to any ordinary person. It makes people think: Shit, I can do that. I can get my hands on a boat, scrape together twenty grand, grow some food, sequester some carbon, and head to bed at night, body tired, soul fulfilled.

GOING VIRAL

I thought the momentum would subside, but it didn't let up. After the first few articles, more and more newspapers, radio shows, TV stations, podcasts, and magazines wanted to cover ocean farming. Grand titles like "The Seas Will Save Us: How an Army of Ocean Farmers Is Starting an Economic Revolution" and "The Underwater Farms That Could Help Stop the Death of Our Oceans" exalted the work we were doing. *Rolling Stone* listed me as one of the twenty-five people shaping the future, with my picture ridiculously splashed right next to Elon Musk's. *Time* magazine came out the same week, naming 3D ocean farming as one of the top inventions that year. Even my Bengali father-in-law, who loves following tech innovations, was no longer bashful about having a fisherman for a son-in-law.

It was a dizzying ride. Part of me was enjoying it, but another part felt unmoored. One day, I was a regular guy, head down, farming my small plot of water; then, suddenly, I was doing interviews and answering emails all day. I was increasingly living a dual life. On the water, I was a worker, covered in mud, with tiny shrimp crusting my beard. I was breaking ice, tumbling oysters, pulling mussel socks, fixing my outboards. But the long days out on the water, tending my farm with my shellfish-eating puppy aboard—they were getting fewer and further between.

The story, and the level of hype, was also getting increasingly out of whack with reality. Two networks approached me to star in a reality show. A theater troupe even turned me into a puppet for their historical play about the Atlantic Ocean; to this day, that puppet lives slouched in a corner of our office, terrifying unsuspecting houseguests. I was being lifted above my station and rubbing elbows with people way above my pay grade. Tamanna was skeptical, to say the least; she said she'd married an ocean farmer, not a media darling. When I was asked to be on a *New York Times* panel about the future of food, though, the first thing she wanted to know was if she could come. And I'm glad she did—she kept me sane, even if she was happily sipping a "compost cocktail."

One day, I got a phone call: "Mr. Smith, good morning." I forget his name, but remember he talked fast. "I just read the article about you in this morning's *Wall Street Journal*. I run a private jet company here in Manhattan."

What the fuck? "Clearly, your business is about to take off, and when you're ready to begin leasing, or getting ready to buy, we're the top private executive airline in the country, and we can get you what you need."

I couldn't even process what he was saying. "Okay," I said, "I don't think I'm ready." I hung up.

Another time, I got a call from the assistant to *Star Trek*'s Captain James T. Kirk. Honest, hand-to-my-heart true.

"Hello?"

"Hi, I'm calling on behalf of William Shatner. He'd like to collaborate with you." Christ, I didn't need a plane, but the captain of the Starship *Enterprise* could surely sell some kelp.

"Wow, cool. Shatner—I'd be way into that. What kind of collaboration are you thinking about?"

"Billboards." Huh?

"Mr. Shatner would like to offer you the opportunity to help market your business."

"How does that work?" I wasn't getting the drift.

"You pay Mr. Shatner a reasonable fee, and you can use his picture to advertise. Many clients find billboards a good way to maximize the impact of Mr. Shatner's fame."

Needless to say, *Star Trek*–themed kelp billboards never materialized on Route 95. But it was indicative of how crazy the shit had gotten. Maybe this was just the way of the world now, social media hungry for the next meme, reporters scanning the interweb for any blip of hopeful content.

Back on the docks, my friends were pleasantly unimpressed with my newfound "fame." It was embarrassing showing up with TV cameras following me. But I thought this was what was needed to keep making kelp the new kale. When I won the Buckminster Fuller Award, which came with prize money, one guy remarked: "Just sounds like a lot of fucking paperwork to me. Now they have their hands in your pockets." The message hit home: Don't be an asshole. Keep your roots.

Though I was feeling a little queasy about all this attention, it sure was helping "take the weed out of seaweed." And I was getting better and better at telling my story. I hated speaking in front of people—I still get sweaty shakes and nausea before any event—but it was moving the needle. Kelp was trending on all the food blogs, and seaweed products were breaking out of the vegan food co-ops onto national retail store shelves. The retail chain Target pushed seaweed cosmetics brands out to the center aisle, Naked Juice rolled out a Sea Greens Delight smoothie, and Akua's kelp and mushroom jerky became all the rage. Two Roads Brewing Company even had kelp beer on tap!

Soon people began calling me the "kelp king." But I didn't want to be king of anything. So what the fuck did I want? Thousands of people were eager to get involved; it was looking like a movement was forming around the vision for restorative ocean farming. But I couldn't shake my growing uneasiness. On the road for event after event to spread the gospel, I'd find myself alone in a hotel room, ordering room service on someone else's dime, missing the stink of low tides, the rhythm of hauling cages, even missing the body ache from a good, honest day's work. In my zeal to sell seaweed to Americans, I was being pulled off the water, and I knew if I didn't change course this uneasiness would morph into personal darkness.

I was too central to this story. Too much of the press and attention focused on me rather than the work. "Storied food" was all the rage in the marketing sector, meaning reconnecting consumers to their food by way of farm-to-table stories, often about the daily lives of their local farmers. So I told my tale every chance I got, learning to hit the right notes to move hearts and minds, my journey of redemption and purpose through meaningful work, a fisherman trying to save our seas. It came from an authentic place, but over time it became rote. Driving to events together, Ron, my first mate, and Tamanna would taunt me with the talking points they'd heard so many times.

"My story is one of ecological redemption," Ron would start.

"Quit it!" I'd yell.

"I dropped out of high school at fourteen, and headed out to sea . . ." Tamanna would continue. She and Ron thought it was hilarious to watch me squirm as they parroted my lines.

"I used to be a fisherman pillaging the seas. . . ."

"But now I'm a restorative ocean farmer, growing shellfish and seaweed. . . ."

They'd trade off like that, laughing and laughing as I cringed at hearing my own set lines. An hour later, I'd be on some panel, repeating them back, word for word.

Somehow, without intention, my personal dream had taken on a life of its own. I was just a guy who had stumbled on something that worked. But now the endless stream of emails, calls, interviews, farm tours consumed all my time and attention, and my farm was suffering. I had to put the CSF on hold—just for one season, I said, but it turned into two, then three. At night, waves of alienation swept over me. I couldn't sleep. After another day telling the same story to yet another journalist, I'd be hyped up and on edge, and I began hating the role I had to play. I was a man living two lives, and it wasn't working. When I was given an award by the Clinton Foundation, I accidentally arrived to meet the former president with a ten-inch fillet knife stashed in my bag.

PASSING THE BATON

The time had come to find a way to harness all this excitement. Over the years, I had built a deep relationship with the folks at the Yale Sustainable Food Program, including Jacqueline Munno, programs manager, who had become a good friend. She pulled me aside after an evening of clam pizza (which of course I couldn't eat, because of my new shellfish allergy) and said, "Bren, it's time for you to share and replicate what you are doing." I spent a few weeks mulling over Jacquie's nudge. Up until that point, I had never thought much about expanding my work. I had a hyperlocal focus, and was more concerned with trying to figure out how, by hook or by crook, to pull a living out of the sea. An unrealistic seed began to sprout: maybe we could channel all this momentum into building a national network of farms, with Thimble Island Ocean Farm as the training ground for the next generation of 3D farmers. Lisa Holmes, a friend and leader in the Connecticut food scene, suggested that a nonprofit model might be a good route. GreenWave was born.

Immediately I realized I was in over my head. I didn't know shit about building a training program, hiring staff, or even filing the right IRS paperwork. I had no idea how to pull this off. Lisa told me I was going to have to learn how to raise money from rich folks. Turned out I sucked at that, too. My first meeting was with a Fairfield County investment tycoon who had reached out during the Kickstarter campaign. It was a total loss—literally. I met him and his wife at a posh restaurant in Westport, ground zero for folks with deep pockets. Out of the gate, I had my knuckles rapped, told that the rich were offended by being calling "rich" and preferred the term "high-net-worth individual." Rolls right off the tongue.

Fire-hosed by drinks and food, I tried to muster the courage to ask for a GreenWave donation. I was nervous about asking for money. How much—a hundred bucks? Five hundred? Five thousand? I had no idea. By the end of the meal, I still hadn't done the ask. Then the dinner check came. My Newfoundland instincts kicked in, and I lunged for it. In my world, dinners always ended in the scrum of guests' fighting to pay the bill, a ritual played out countless times by family and friends. But I lunged and he didn't. I paid for a nearly five-hundred-dollar meal. I must be the only person who's ever gone to dinner with a prospective donor and lost money on the deal.

I was never going to be an organizational man. Who was I

fooling? Christ, I like an eleven-in-the-morning Pabst Blue Ribbon on the way home from the farm. I swear bloody murder at my deckhands when they fuck up. I was a lawsuit waiting to happen. I needed help. In a stroke of luck, right around this time, I was getting nagging phone calls from a kid named Brendan Coffey. He kept calling, leaving message after message. I wasn't calling him back, but the kid wouldn't fucking let up. One day, I finally picked up, annoyed: "What do you want?" He launched into his love for the oceans, his passion for the cause, and his desire to help. He called himself a "social entrepreneur," lived in Manhattan, and had founded the global Fishackathon with the U.S. State Department. He came from a different planet, for sure, but maybe that's what I needed to get this job done.

Brendan got to work building a website, registering the nonprofit, building a database, and setting up a board of directors. Mining the social-entrepreneur world, he applied for and got us fellowships with Echoing Green and Ashoka, organizations designed to help projects like mine get off the ground. These early sources of funding and mentorship were critical to launching GreenWave. The kid kicked ass but was just as inexperienced as me at running an organization. I needed someone who could take the helm and turn GreenWave from fledgling concept into a functioning organization.

Along came Emily Stengel. At first I thought it was a bad fit: she had just finished grad school and previously ran a high-end Manhattan catering company for the famed chef Mary Cleaver. Didn't know shit about the oceans, seafood, or aquaculture. Landlubber, through and through. When she came to my house for an interview, I learned about her work. In school, she'd toured the country and interviewed more than two hundred land-based farmers, asking them about health care, wages, housing costs. She learned firsthand that not one of them was making ends meet. And the catering company wasn't some little boutique gig—she ran a crew of ninety people in the unforgiving wilds of the Hamptons; she had serious organizational chops. But, most important, I sensed she was good people. Maybe it came from the mixture of her Philly roots and the hectic grind of the Manhattan catering scene, but she had an unflappable, no-bullshit temperament combined with empathy for the struggles of everyday people.

So I hired Emily, and we set up shop at the Quinnipiac River Marina, to this day one of the most productive oyster grounds on the East Coast. We began putting our plans into action, starting with the farmer training program, which included hands-on education, free seed for two years, a start-up grant, permitting support, and winter gear donated by Patagonia. Emily hired Kendall Barbery, who had fished six salmon seasons in Bristol Bay, Alaska, and graduated from the Yale School of Forestry & Environmental Studies, to be our programs manager. We started working to develop policy frameworks for ocean farming in New York and California, bringing aboard Karen Gray, who had worked for the California Coastal Commission, to launch the GreenWave Reef in California and help farmers navigate the most complex permitting regime in the country. We were building a large-scale seaweed hatchery in one of the poorest neighborhoods in New England, while incubating new kelp-centered companies to expand markets for farmers' crops. With a competent, passionate crew beside me, I was amazed how quickly ideas came to fruition.

It took a couple years and some trial and error, but we ended up with an awesome team. David Henry—like Emily, a fish out of water at GreenWave—heralded from Texas and had spent the last decade fund-raising for an organization fighting child trafficking. He came on as director of development. Liana Coviello was hired as our executive assistant, having worked for many years on the executive team of the famed Woods Hole Marine Biological Laboratory. And of course, Ron Gautreau, my long-time first mate, stayed on as our senior farmer trainer.

GreenWave was pulling all the pieces together, turning the twenty-five-farm "reef model" into a reality. And it was finally reconciling my two hearts and two minds. I'd found a way to live both on the water and in the world of ideas. I was learning way above my pay grade—about new models of economic development, carbon offset mechanisms, traceability standards, underwater sensor technologies. It was exhausting—and exhilarating.

WHO FARMS MATTERS

One of GreenWave's first new farmers was Dave Blainey, an eleventh-generation fisherman out of Rhode Island, who, like me, fished from the Georges Banks to the Bering Sea. In his late fifties, he was deadly serious; the tooth of a shark he had hooked hung from his neck.

Coming out of the wild fishery, Dave had long been suspicious of aquaculture as an encroachment threat to traditional fishing grounds. But with the fish drying up and ever fewer commercial boats on the water, he was also not blind to the changes. He'd been keeping an eye on my work, and, the more he mulled it, the more ocean farming was starting to make sense.

When we first met, he told me that part of the attraction was simply lifestyle: "You know, as I got older, I was more interested in staying a bit closer to home." Plus, the more he thought about it, seaweed wasn't as foreign as it seemed: there's a seaweed tradition running through his hometown. When he was a child, Dave remembers, local farmers would walk down to the beach to collect seaweed to use as fertilizer on their fields, and the walls of some of the older buildings in his town are packed with seaweed, commonly used as free insulation in the days before fiberglass and foam. And he liked the old-school feel of it, with "just about everything done by hand. It's a way to continue working on the water, just in a more relaxed way."

After doing some training with Dave—which was lightning-fast, because he already had many of the skills—he quickly became a GreenWave leader in Rhode Island, as both a spokesman and a trainer for other farmers. He took to seaweed farming, saying, "It turns out to be really satisfying to watch something grow." Only a few years later, he's been working with U.S. Senator Sheldon Whitehouse of Rhode Island to secure funding for ocean farming as a climate-mitigation strategy. He'd better be careful: on his current track, he'll find himself drafted as the first ocean farmer running for elected office in America.

But it wasn't just former fishermen who were attracted to the GreenWave model. Young women soon began showing up in droves. This was unexpected and groundbreaking. Men had dominated the marine space for hundreds of years. What if women could emerge as the leaders in building this new blue economy?

Jill Pegnataro was one of the first to come knocking. Her family had once owned Pegnataro's, a much-loved local supermarket, before it shut down in the 1980s. Both charismatic and tough, she became interested in growing lobsters at an early age, when she attended the Sound School, a public vocational aquaculture high school in New Haven.

"I loved them for their ferocity," she said. "Even the juvenile lobsters, no bigger than a pinkie nail, need to be separated or they will fight to the death. I was fascinated by their persistence." She continued to study lobsters in college, and emerged as a full-throated environmentalist. But after she graduated, the jobs weren't there. Like so many of her generation, Jill ended up in the shit-job economy. As she tells the story, after a string of internships, "I looked for a permanent job, but the economy had other plans. I took temporary administrative jobs and struggled in each of them. I feared that I was never going to find a job in my field, doing what I loved."

Then she heard about ocean farming and "felt sparks of hope." Later, after learning to farm, she told me she was proud and thankful that she had found a way, in her words, "to do the environmental work that gives my life meaning and allows me to make a difference." Jill eventually became GreenWave's farm manager, and is well on her way to becoming a national ocean-farming expert.

Up in my old stomping grounds of Alaska, Tomi Marsh, a fisher who has been captaining her own boat for decades, has emerged as a leading voice and innovator. When she was twenty-six, she bought her first boat, the F/V *Savage,* and crabbed, tendered, and long-lined from the Bering Sea to the West Coast. Tomi told me she first got excited about ocean farming at the age of five, when she watched *Twenty Thousand Leagues Under the Sea:* "Watching Captain Nemo show his guests his underwater farm and reading Verne's descriptions of the foods and his spot-on views of ecological farming inspired the five-year-old me."

After forty years of working on the ocean, Tomi is now farming like Captain Nemo. She started OceansAlaska, located in Ketchikan, to make 3D ocean farming a reality in Alaska. With the help of a diverse team, she built a local shellfish and seaweed hatchery and is now working with a wide array of fishermen and native Alaskans to diversify Alaska's fishery. According to Tomi, "I knew I wanted to take GreenWave's ideas of vertical farming and the attendant ecosystem services and help to make that model a reality in Alaska; a modern-day Captain Nemo sea farm!" After a mere two years, Alaskan farmers are now growing hundreds of thousands of pounds of kelp and selling to new Alaskan seaweed companies like Blue Evolution.

Through GreenWave, I was quickly working myself out of a job, as new farmers emerged every day, exactly what I'd hoped for. They now come from all walks of life. Jay Douglas, a former marine who has done tours in Iraq and Afghanistan, and his wife, Suzie Flores, a former bartender turned editor, now farm in Stonington, Connecticut. Mark Kelleher, a retired teacher, is farming in Nantucket Sound. Out in California, where Green-Wave teamed up with the Port of San Diego and Hog Island Oyster Company to support new farmers, women are once again leading the charge. In the Bay Area, Tessa Emmer, Catherine O'Hare, and Avery Resor planted the first commercial seaweed farm in California, quickly followed by Torre Polizzi and Leslie Booher in San Diego.

By 2017, GreenWave was fielding requests to start farms in every coastal state and province in North America and twenty countries around the world, and our website crashed under the weight of a million visits in one month. A tsunami had hit shore.

CHAPTER 17

Swimming with Sharks

J ust as things were heating up at GreenWave, I got a cryptic email, "If you want to get serious about this, I'm an investor, come see me." I liked the cloak-and-dagger tenor of the message and figured I didn't have anything to lose, so I drove down to Greenwich and had my first-ever meeting with an investment banker. Swank office: leather couches, water with cucumber. I just came as myself—stained work clothes and Whalers hat; I didn't have much to prove to the Wall Street crowd.

He invited me in, and I was surprised not to meet the blood-thirsty finance shark I had expected. He was gracious and thoughtful. He had made a shitload of money in the markets and was hungry to recast himself as an environmental super-hero. "I've done well; now I want to shift gears," he told me. "I want to help." In the coming years, I'd hear variations on this from dozens of investors: it was time to "make a difference," "give back to the world," "do something game-changing," "make my mark." These were a new species of human I hadn't met before, but I liked the redemptive tones.

I was halfway through telling my fisherman-to-farmer story when he interrupted me: "What if I invested two million dollars right now?"

I had no idea what to say. What kind of guy meets someone the first time and asks if they want two million dollars? I stammered—no plan prepared. I told him I'd build ten farms a year. He shifted from eager to unimpressed. My vision was too slow and modest, my little nonprofit a distraction. He wanted to hear about scale—thousands of farms, billions of dollars. He was looking for bigger fish to fry.

A few weeks later, I headed down to Bridgeport, Connecticut, for a dinner hosted by the New England Food Summit. It was a typical gathering of the food-activist crowd: funders, well-heeled college kids, a few politicians, "slow food" investors, and a smattering of farmers and immigrant-rights groups and food-justice organizations.

Across the table, I heard the voice of a woman who had the gruff air of a fisherman. I caught a few words: "ocean farming" and "kelp." I jumped in, asking her to back up and explain. Her name was Paula, and she was from Rhode Island. She ran an oyster farm, and her husband was a shrimper. They had been approached by an "investor type" who talked a mile a minute about how he wanted to "revolutionize and scale the industry" and take Rhode Island to the next level by pouring millions into 3D ocean farms. He told Paula he'd pay for all the farm gear and buy everything they grew. I heard the inflections of Wall Street. I asked for his number and called him up; turns out I was right. Late twenties, backed by a gaggle of unnamed investors eager to "go green while putting green in the bank."

I invited Paula; her husband, Adam; and their buddy Greg

down to the farm. These were my kind of folks. A swill of swagger, salt, and suspicion, still bleary-eyed from pulling an all-nighter on the shrimp boat. I showed them the farm, how straightforward it was to build and run. We talked about the big money headed their way—hundreds of thousands of Wall Street dollars on the table to set up farms. I told them it was their decision and I was going to stay out of it, but wanted them to get a clear picture of what they were getting into.

GreenWave had access to a team of New York lawyers, the firm Gibson Dunn, who had donated a thousand hours of free legal help to GreenWave. We asked them to look over the contracts being offered and to advise the budding ocean farmers. I had no idea how the relationship would play out. Adam reported back proudly: "Thank fucking God, Gibson Dunn is on our side—those people are killers." He was proud that they were sitting on his side of the negotiating table. As the lawyers dug into the details, it turned out that the proposed contract was typical Wall Street fare. Investors would pay for farmers' gear, and supply seed, but lock them into a long-term contract at below-market prices. Adam had been fishing his whole life—often selling to Chinese global buyers—and knew this game. Once the lawyers unpacked the opaque contract terms, Adam and Paula walked away. Good instincts.

It felt like a win, but it wasn't. Though I didn't know it then, this marked the beginning of my time swimming with sharks, which would turn out to be one of the worst periods of my life.

SURFING THE "NEW ECONOMY"

Word was out: There were big bucks to make at sea and farmers were chum in the water. Corporations began realizing they could move in early and dominate a new agricultural space with unlimited growth potential. Sharks began circling, calling, emailing, showing up at events. A Prince from Qatar. Head of economic innovation at the White House. Breathless stories came out in *Forbes,* the *Financial Times, The Wall Street Journal.* Others were promising millions in Bitcoin investments. I heard rumors of a Saudi arms dealer who was trying to get his hands on a California lease. Tax lawyers were salivating at the prospect of using farms in international waters as tax havens. One expert interviewed by *Business Insider* captured the mood: "If I could buy kelp futures, I would."

At first, I tried to keep my head down and ignore it all—my job was farming, not finance—but I faced a huge hurdle. Green-Wave was training more and more farmers, which meant more and more kelp and shellfish hitting the docks. They needed a way to process and sell all these crops. When it was just me, my small processing facility in a rented warehouse space set up with discount restaurant gear worked well enough. And farming so near to New York City meant I had quick and easy access to high-end markets. But we were quickly entering new terrain, requiring large-scale processing and distribution. Shellfish didn't require much processing infrastructure. Indeed, the kick-ass thing about oysters, clams, and mussels is that they come already packaged, in their own shells. Just pull 'em out of the water, clean them off, and they're ready to sell.

But kelp was another matter. It has a hyper-short shelf life—less than half an hour. It hates oxygen, and begins to shrivel and rot soon after being hauled out of the water. The solution was to get it to a processing facility to blanch and freeze as soon as possible, preferably within twelve hours of harvest. The prospect of having hundreds of thousands of pounds of kelp needing to be processed during a mere five-week harvest period scared the shit out of me. Farms were cheap and easy to build, but once we hit land, the economics changed: we needed buildings, processing machinery, freezers, trucks. This could cost millions of dollars. Fuck.

On my journeys, I repeatedly heard rumors of a new emerging sector of finance that had shape-shifted into self-described "impact investors" in the wake of the 2008 financial crisis. As I started making the rounds to meet this new breed of shark, I quickly felt like a study-abroad student arriving without a *Lonely Planet* guide. The first meetings were daunting: they spoke aggressively and used coded financial language. At their posh dinners, I'd feel so uncomfortable that I'd just push my food around the plate, nerves squashing my appetite.

But as I sat, sipped, and listened, I learned that the one-percenters are as dazed and confused as the rest of America. The compass that had guided their families to wealth over the last hundred years had lost its bearing. Like the rest of us, they are running scared as the wildfires burn, soils shrivel, and seas relentlessly rise. I was surprised to hear the same refrains sung by those of us at the bottom of the food chain: how climate solutions need to protect both people and the planet; how inequality and social justice are central, not side, issues; how farmers deserve a more equitable share of the economic pie. Plus, they had the added worry about the pitchfork-wielding masses gathering outside their office suites.

We seemed to have common cause, but my feelings of alienation ran deep. At one dinner with potential investors in a penthouse suite overlooking the Hudson River, the host turned to me mid-meal, prepared by his private chef, and said, "Bren, tell us an exciting story about being a fisherman!" Everyone turned to me. I suddenly realized I was the entertainment, the exotic shiny toy of the evening. I wasn't being welcomed into a community; I was on temporary display until they brought in a new exhibit.

So I told them the story about falling asleep on the toilet in the middle of a brutal shift and waking to find that I had shit myself. It was the first story that came to mind, but I suspect part of me was trying to spoil the meal. One woman yelled "Grossss!" Then they all burst out laughing, pleased. I had performed as anticipated, providing just the kind of off-color behavior they expected from a fisherman. They continued with their meal, chatting about their other "disruptive investments."

At another dinner, I excused myself to go to the bathroom before making my escape. The heavens played the cruelest of jokes. The bathroom, larger than my old Airstream trailer, had a crazy-looking toilet—all buttons and knobs and arrows. I took a leak and tried to find the button for "flush." I'm sure it was there, somewhere obvious to others, but I couldn't find it, blinded by thoughts: Fucking kidding me, I can't even flush these people's toilets. What the fuck am I doing here? I ducked out, leaving the piss bubbles floating.

I'd never belong, but I got a peek inside, and it became clear that while they were indeed scared of the darkening future, the vast majority of "impact investors" hadn't come to grips with the reality that protecting the planet was going to hurt their bottom line—that the "impact" they wished for was going to shave off a significant portion of their profits. One time, a group of investors flew in for the day; I had been talking to them for months, feeling I had finally found the right partners. They were pumped up, hitting all the sustainability high notes, clearly excited to have found an opportunity to use their money for good. But the whole meeting screeched to a halt when I asked a simple question, "What are you expecting for a return?" Answer: 20 to 30 percent in five years. Nothing had changed. They expected to match the profit margins they had been raking in for decades. One even said he anticipated making a higher return on his sustainable businesses portfolio because he saw high consumer demand for "ethical" products.

In 2018, Jeremy Grantham, the hedge fund titan famous for predicting the last two major market bubbles, told a packed room of investors that capitalism is killing the planet and needs to change. "We deforest the land, we degrade our soils, we pollute and overuse our water and we treat air like an open sewer, and we do it all off the balance sheet. Capitalism and mainstream economics simply cannot deal with these problems."

In other words, for decades, corporations have been making money by shoving the costs associated with polluting our planet onto the rest of society. They pump carbon into the atmosphere and oceans, and then we, the taxpayers, have to pay to clean it up. These captains of Wall Street are peddling their own brand of climate denial. Or they pay employees crap wages and no benefits, so even those working full-time need "government handouts" to afford housing and health care. We're all paying so the 1 percent can haul in that 30 percent return.

I'd be damned if I was going to be part of turning restorative farming into a carbon copy of vertically integrated industrial aquaculture and agriculture, in which profits accumulated for the captains of industry while fishermen, farmers, and the planet bore the burden. My tour of Wall Street convinced me that I had to find allies who weren't content with business-as-usual while the world burns.

BLOOD IN THE WATER

Mike Shim to the rescue. I had known him for several years as the founder and CEO of Ocean's Halo, a hugely successful boutique seaweed company. He'd begun to crack the nut of creating a wholesome, tasty, value-added brand. But now, in his own words, he "wanted to move to the boring part of the problem." It was a humble way of saying that creating a new brand was the fun part—wrapping a story around seaweed, creating compelling marketing materials, and taste testing—but processing millions of pounds of kelp? Not so much. This work required figuring out blanching systems, drying technologies, transport logistics. Mike suggested bringing in his friend Norman Villarina as a partner.

We were from different planets, but maybe that's what makes for a good stone soup. Mike was the start-up guy who knew how to build companies from the ground up. Norman was a self-described "shark that keeps the sharks away," aka a "fuzzy shark." He'd cut his teeth on Wall Street, but now was running an investment fund supporting everything from solar farms to almond-grower co-ops. These were the species of sharks I needed.

They flew in for a sit-down to explore a potential partnership, and we met at The Study at Yale, a hotel and restaurant that I had never had the money to set foot in.

Norman explained that, to raise money, they'd need to conduct a background check on me, because investors demanded security checks before writing checks. It would be an exhaustive five-week process that would cost ten thousand dollars; Norman had one done on himself every two years. Weird.

"Not happening," I said.

"Why not? Is there something you're hiding?" Mike asked.

"Listen, I'm hiding nothing: I've been arrested for assault and battery, assault on a police officer. I dealt drugs for years. You don't have to blow ten thousand dollars. Just ask me what you want to know."

The blood drained from Mike's face. Norman looked confused and was surely thinking: What has Mike gotten me into?

"But, um," Norman stumbled, "you're not gonna get arrested or anything again, right?"

"No, probably not. I've got a family now. But those were some of the best days of my life. I miss them all the time."

"But investors . . ." Norman continued.

"If anyone needs a background check, it's the investors," I interrupted. "If anyone's broken the law lately, it's them."

We went back and forth like that for a while—this is what happens when you try to climb into bed with a fisherman. In the end, we skipped the background checks. I may not have been business-school material, but I held a couple key pieces of the puzzle. I had built a well-known sea greens brand and been selling kelp for years. I had a long roster of contacts and knew the farming sector inside and out. After some deliberation, we all decided to give the partnership a try: Sea Greens Farms was born.

Early on, we had vision-driven conversations. We talked about farmers' being co-owners of microprocessing plants spread around the country, helping set up co-ops so growers could collectively bargain for a higher share of the profits, hiring former felons and paying all workers living wages. Things felt as though they were falling into place: the vision of the GreenWave Reef, finally becoming a reality, built by people with similar values of shared prosperity up and down the value chain. Our dreams seemed aligned, and I breathed a sigh of relief. Finally, some financial experts to shoulder the burden.

First order of business was hiring a powerhouse general manager to get the processing plant up and running. With more than a decade of spadework, there were huge buyers now poised to purchase kelp—Google, Whole Foods, even Compass Group, the largest food logistics company in the United States. We had to deliver.

We got a lot of responses, but one guy stood out. On paper, he appeared a godsend. A lawyer by trade, he had owned a seafood-processing plant, was an early importer of farmed tuna from Japan, and claimed to have launched eleven new start-ups in the food industry. No other applicant even came close to having the skills we needed to set up Sea Greens Farms' operations. But when we met, despite the stellar qualifications, there was just something off about him, though I couldn't put my finger on it. After hemming and hawing for a few weeks, I ignored my gut instinct and took the bait. I was exhausted and overwhelmed with work and the demands of now two separate organizations, and we had to have someone aboard before the next harvest season.

I hired a shark and he was there to feed. Over the next nine months we let him take the lead, trusting his commitment and expertise. But then we got the heads-up from a few of the farmers we worked with that he was secretly setting up a competing business. While overall as a farmer I viewed growing competition on the buyer side as a good thing—more companies fighting to buy our crops meant higher prices at the docks—it was still a shock that someone would try to do it from the inside. Just a shitty, dishonest, low-road thing to do on Sea Greens Farms' dime. Sensibly, Mike fired the shark, and we decided to simply chalk it up to growing pains and move on. There was important work to do. But instead of walking away and continuing to set up his own company, the shark turned around and filed a lawsuit against Sea Greens Farms for hundreds of thousands of dollars. He knew our financial situation well: we were a tiny start-up that would be dragged under by the cost of a lawsuit. He knew we didn't have that kind of money; I began to suspect his real goal was to drive us under.

The more I learned, the more I became convinced that this guy had been following a preestablished playbook: to prey on start-ups by bleeding us dry. Should have done that ten-thousand-dollar background check, because it would have revealed this shark to be a serial litigator who had been involved in close to twenty lawsuits since the 1990s.

Mike and Norman stepped up and took on the fight to save Sea Greens Farms, refusing to bow down to his tactics. For my part, I tried to ignore it, stay steadfast in my own work, but I could feel myself rotting from the inside. Month after month, the lawsuit dragged on. I was stunned that someone else's greed might so easily derail everything so many of us had worked on.

After months of sleepless nights, I decided to call it quits. Mike and Norman had the guts to fight it out in the never-ending court battles, but I didn't. I walked away from Sea Greens Farms, and pushed for my equity stake—I owned a third of the company—to go to a reserve profit-sharing fund for farmers, but that idea got shut down curtly as too starry-eyed. So I just handed all of it over to Norman and Mike. I didn't give a shit; I was never out to get rich or run a big company. My goal was to grow more kelp, soak up some carbon, and have more farmers growing good food and making a decent living.

The year had taken its toll. I had sacrificed my cherished days on the farm for bottomless lawsuits, mind-numbing conference calls, and just general hamster-wheel shit work that had my body wilting from lack of use. I was in a dark place, questioning everything I had built. At work, I was in a bitter mood, even questioning the purpose of GreenWave. At home, I wasn't much better. I felt the dream of farmer co-ops, profit-sharing processing plants, and food justice slipping away.

The more I reflected, the more I realized that the story of Sea Greens Farms wasn't a morality tale about a bad apple abusing the legal system to get ahead. It was the story of my own failure to keep a budding industry from devolving into business-as-usual. I thought I could help change the rules, but I was wrong. After a year swimming with both fuzzy and bloodthirsty sharks, I didn't see anything "new" in this "new economy." What was the point of building something from the ground up if it was destined to be taken over by sharks? It felt like they were trying to tear the vision to shreds, and I was just a pawn in their winner-take-all game. It hurt more than I thought it could.

If this was the world we were living in, I didn't want any fucking part of it.

CHAPTER 18

Newfoundland, Take Me Home

When lost, retrace your steps.

Tamanna decided it was time for me to get the fuck out of town and head back to where I had first crawled from the sea. I hadn't stepped foot on the Rock for more than a decade. With my accent gone and patience tuned to an American frequency, it showed. To the island, I was now a "come from away," someone who called the mainland home.

But I had nowhere else to go. A few months earlier, I had driven back up to Lynn, Massachusetts, to see my old port of call. Not a single commercial boat was moored in the harbor. My mom's beat-up old house had been fixed up. Even the public housing I had lived in with my teen girlfriend had been torn down. There was nothing for me there now.

So we repurposed my orange work truck into a camper, building a bed in the back, drawers for kitchen supplies, food, and piles of sweaters, hats, gloves, down coats, and rain gear. Tamanna christened her "Clementine." We planned an eighteen-day road trip to the Rock, living out of the back with Juniper, our eleven-month-old Newfoundland, already a ninety-pound giant.

We drove for two and a half days just to get to the ferry, a sixteen-hour overnight ride from North Sydney, Nova Scotia, to Argentia, Newfoundland. Tamanna, a bit seasick, went to bed early, while I sat in the ferry bar, mind still mired in the problems back in the United States. But I was also worried about what might have changed in Newfoundland. I'd seen the ports of New England turned into tourist destinations, with more vacationers than fishermen in the bars. I don't think of myself as someone who clings to the past; everything changes, and I'm good with that. But it's the hollowing out without replacement that worries me. I didn't know what I'd find, what my guts would feel.

I got up at 4:00 a.m., after a few hours of sleep, to check on Juniper, let her take a piss after an unhappy trip in the ship's kennel. It was still dark, and I caught sight of the lights along the southern shore of my long-ago home, clusters of outports still only accessible by boat. As the sun rose, so did the land—the towering rock cliffs and barren hills stretching to the horizon. I warmed a little.

We drove off the ferry at 8:30 a.m. and started our journey along the water's edge of the Avalon Peninsula. Tamanna was thrilled by the colors of the outport homes—reds, greens, pinks—nestled into the coves. I was surprised to see so many trawlers, crabbers, scallop boats coming and going, and processing plants up and running. Last time I was back, the island was still reeling from the 1992 closure of the fishery, the docks left for dead.

We pulled Clementine into an outport named Branch, parking on a pebbled beach to make a pot of coffee and give Tamanna a chance to do some world-class beachcombing. Smashed lobster traps, scraps of nets and twine, gloves, bird and whale bones. Never thought I'd bring a wife from Los Angeles to the shores of Newfoundland. I searched for sea glass as a gift, but couldn't find any, the world having switched fully to plastic bottles and cans since the last time I'd visited. Juniper found a bird carcass and paraded it just out of our reach.

A pickup truck that was driving by took a sudden sharp turn off the road when the driver spotted us. He pulled over to see what we were up to, shouting: "Beautiful dog. Small, though, for a Newfoundland—how old?"

"Eleven months," I shouted back, heading over to have myself a good chat.

"Ah, she's a sook," he says, using the Newfoundland term for a needy, loving beast; he looked like he wanted to keep her for his own. He asked where we were from and how we liked it so far, and reported we'd just missed the capelin run, that favorite time of mine, when waves of tiny, oily fish roll onto the beach to spawn. Tamanna could barely understand his distinct outport accent.

"Would have been something to show my wife, a run of capelin," I said.

"Do you eat it?" he wanted to know.

"Sure," I said. "Used to as a kid, at least. Fry it up."

He turned to Tamanna, wanted to know if she'd ever tried it. She shook her head. He paused, looked back and forth between us, then asked, "You want a feed of capelin?"

"Sure," she says, "I love any fish." Bengalis and Newfs belong together.

"Be right back," he says. We were a bit confused, but he made a quick turn and sped away in his truck. He returned quickly, in less than five minutes, and rolled up holding a Ziploc of frozen capelin, a fish I hadn't laid eyes on in decades. Tamanna took the bag, amazed that our first stop had yielded free fish from a stranger.

Like many Newfs, he talked fast and swerved from topic to topic, an essential skill for keeping a chat going for hours with friends and strangers alike.

"I used to travel a lot," he says. "Drive up to St. John's, sometimes to Canada."

On the drive, I had briefed Tamanna on many Newfoundlanders' insistence on calling the mainland "Canada," as if our province was still its own republic.

"But I was drinking too much and getting in too much trouble. So I quit."

"You quit drinking?" I asked.

"No, b'y!" he exclaimed, appearing offended. "I quit driving. I still loves me beer."

There it was, the art of Newfoundland conversation. The love of wordplay, the importance of wit. The collapse of a hundred years of political struggle into a precise use of "Canada." A "feed" of capelin, and dogs as "sooks." And that generosity of spirit that searched out every opportunity to give a gift. Newfoundland is the poorest province in Canada but gives more to charity than any other; this fact sums up pretty much everything you need to know about Newfs. I'd forgotten this approach to living. The radical, aggressive hospitality of the people—this is the culture I had been raised in. I suddenly felt a world away from the hungry sharks.

We drove endless winding roads along rugged coasts, through windswept barrens, parked atop severe cliffs with waves crashing hundreds of feet below. We'd find a place that suited us, and stay awhile. We sat quietly, made simple meals on our camp stove. We took long, meandering walks through bogs where Tamanna reveled in the late-summer Newfoundland tradition of berry picking, something I had done my entire childhood with my family. Tamanna was in awe, constantly commenting on how dramatic it all was, how it was as beautiful as the California coast—the highest compliment, coming from her. Seeing her fall in love with my home let me see it through new eyes, too. This was the place that made me, a wild and restless and giving place. The troubles of the past year slowly started to fade.

Each little outport had its own distinct character, often with its own dialect. Many of these tiny towns were no more than a collection of ten or twelve houses arrayed around a cove, and some had spent a hundred years in relative isolation until roads made them accessible. I remembered why this place had been a dream for my father, the linguist. I picked up a copy of the *Dictionary of Newfoundland English* and traced his name in the prologue. All my habits of phrase that seemed odd in the States made sense here.

One day, we were parked below a lighthouse when two guys parked their motorcycles next to us to meet Juniper. Craig and Rich were proud "Newfound Riders" and self-declared "outport boys" who had moved to St. John's to work, one as a nurse, the other as a plumber. Their mascot was the Newfoundland dog, and they knelt for photos with our pup. They rode every weekend for charity, raising hundreds of thousands of dollars, often putting out emergency calls on Facebook for their fellow riders to raise quick cash for those dealt unlucky hands. Their latest ride was for the funeral of a little girl who, knowing she would die by the age of nine from a rare form of cancer, raised more than two hundred thousand dollars running a lemonade stand and donated every penny to children who couldn't afford cancer treatment. We chatted for about an hour; before they left, Craig gave me the hat off his own head, a memento.

We continued along the Burin Peninsula and stopped in the town of St. Lawrence to check out the Miners' Memorial Museum. The woman who ran the museum insisted on giving us a personal guided tour—something she does with every visitor. She began: "I want to prepare you—this is a sad story, one that will crush you, but that will peak several times and leave you without sorrow." Tamanna and I looked at each other in silent anticipation; this was going to be good.

Our guide told a tale of how her town turned from fishing to mining in the early years of industrialization. Of fishermen promised a better life underground, digging for fluorspar, used for everything from making aluminum to refining petroleum. She showed us their tools, photos of proud workingmen, and the piano that was played when mining shifts were over.

Then, as promised, the story took a dark turn, as we gazed at murals of mothers in black dresses clinging to daughters with charcoal eyes. Most men died from silicosis in their early forties, leaving widows and children to haunt the streets. Then the story tilted upward, as the women fought back and the men unionized, winning ventilation gear and survivors' benefits, and then back down we went, as the companies falsified medical records to cheat families out of their breadwinners' death benefits. My former labor-organizer wife was openly crying. Our guide appeared used to this—she had a knack for bringing such stories to life as only a Newfoundlander could.

We moved on to the next exhibit. This one was a story I had heard many times at school, about the winter storm in February 1942 that shipwrecked two U.S. naval destroyers. Two hundred and three soldiers died, but coal miners, just getting off their fourteen-hour shift, managed to save 186 men.

This was certainly a heartwarming history of local valor, but it didn't convey the moral of the tale. There were four people of color onboard, three Filipinos and one African American from Mississippi. The Filipinos elected to stay aboard, because they thought they had landed in Iceland, their intended destination, where they knew they could be publicly flogged for interacting with white people. Those three men died. The Mississippi native, Lanier Phillips, decided to take his chances. With the other survivors, he swam ashore and was carried, near frozen to death and soaked in oil, to a house where Violet Pike, the wife of a coal miner, undressed him and began scrubbing off the oil. Violet scrubbed so hard that Lanier began to bleed, until he stopped her, saying, "It's the color of my skin, you can't get it off." Violet, like all the others in the town, had never seen a person of color before.

The U.S. Navy was strictly segregated then, and Lanier was from a town where he could be lynched for being found alone in a white woman's bed. He was certain that once they realized he was black he'd be thrown out, or worse. Instead, the woman spoon-fed him soup, and nursed him back to health.

As Lanier tells it, the moment that changed his life was being invited to sit for breakfast with the family of six. He had never known there were places in the world where a black man could be invited to share a meal as an equal with a white family, and it changed him. He went on to march with Martin Luther King, Jr., and became a leader in the fight to desegregate the navy. He returned to Newfoundland many times before his death, and credits that meal as a pivotal moment in his life.

We weren't just being told a sentimental tale. It was a grounding saga that, in its repeated telling, allowed this small town to understand its place in the modern world. What was once a World War II naval adventure story was recast as a civil rights struggle. The heroism of the miners was important, but the heart of the exhibit was Lanier's story, a proof of the enduring power of empathy as the modern mainland goes dark with division.

As Tamanna and I continued poking our way along the coast, I began to sense that Newfoundlanders were working the same sorcery on their struggling fishery—remaking tradition to fit modern times. When I last visited, every conversation had twisted toward a darkness: young people fleeing the outports, boats beached, work without meaning. Doug Dunn, a fisherman from Renews, captures the thoughts of those days:

> You were walking around that year of the moratorium and you were lost. And nothing you could do about it. Not a thing . . . I was making rock walls that I didn't want to be at because I was supposed to be on the water. You were looking for something and you didn't know what you wanted.

Lost ghosts, hungry to return to the water. I had journeyed to Newfoundland because, although pushed by different forces, I felt that, too.

Near Burin, I stopped by a fish-processing plant to find out what was running through the conveyor belts these days. "Sea cucumbers!" an old salt told me with glee. Unreal.

"What you catching them for?" I asked.

"The Algerians," he said, in full grin and wink. "They use them for aphrodisiacs!"

I suspected the Chinese were the end consumer, but who knows, given the tangled mess of globalization, maybe the Algerians were competing with the Chinese to keep their cocks hard. A few other fishermen gathered around; it turned out they were hanging in there, fishing a bit of everything—halibut, crab, and scallops—but sea cucumbers were the cash crop.

Twenty years after the cod stocks crashed, these fishermen were still hungry, but no longer ghosts. They were tilting their narrative arcs upward, pouring new rum into old bottles. This is something I used to be pretty good at. I had become an ocean farmer out of necessity, not anger. I worked from a place of discovery and trust in the endless bounty of the sea.

SEAWEED ON THE ROCKS

Tamanna heard about a new distillery in Clarke's Beach that was making a buzz, so we headed up there for the day. We walked in and found seaweed plastered on every wall! We learned that their top-selling product was a seaweed gin made with infused dulse that had won two gold medals at the San Francisco World Spirits Competition. No easy feat. The other best-seller was the Gunpowder and Wild Rose Rum, made with Newfoundland kelp. Tamanna found the gin subtle, botanical with just a hint of brine to it, the rum smoky and complex. If you want to be a seaweed farmer when you grow up, I suggest finding a wife who's both a nurse to heal your wounds and a foodie to make your crops sound edible.

We continued on to camp in Grates Cove, on the northern-most tip of the Bay de Verde Peninsula. We parked and slept on the cliffs in the middle of town, and I woke with that end-of-the-world feeling I've felt before only in the Aleutian Islands: as if the earth is flat and you're sitting on the edge.

We walked up into the town and discovered a place called the Open Studio Restaurant, which had been a schoolhouse before the cod stocks crashed. The chalkboard menu listed seafood gumbo, lobster pie, seared scallops, partridgeberry pudding. We had accidentally stumbled into a Newfoundland Cajun restaurant. Huh?

The owner came out of the kitchen and immediately noticed the "Kelp Is the New Kale" shirt I was peacocking. "It's true, kelp *is* the new kale," she said. Her name was Courtney, a Louisiana native who married a local artist; together they fused traditional Newfoundland ingredients into homespun Cajun meals. She shared that she harvested more than a dozen types of seaweed on the coasts, the freshest vegetables on the island. The meal was amazing—fresh, thoughtful, and heartwarming.

It didn't stop there. At the St. John's Farmers' Market, Shawn Dawson, owner of Barking Kettle Farm, was regularly selling out of his gherkinlike kelp pickles. He used a homespun Newfoundland pickling recipe: a brine of vinegar, water, and sugar with ginger, coriander, black pepper, and kelp.

What the fuck? Seaweed gin, Cajun sea vegetables, and pickled kelp? Seaweed was everywhere! We even heard a tale of a fisherman who lived on an island with his sheep and ate only seaweed. Trust Newfoundlanders to be ahead of the curve.

Day after day, Tamanna, Juniper, and I explored and listened. I had left Newfoundland long ago, and was never to belong to the Rock again, but this bottom-up innovation by people from the poorest province in Canada was lifting my darkness. It wasn't the sharks that were leading the way. It was regular folks, armed with uncrushable hope and blue-collar drive. Tamanna noticed my eyes brightening. I felt my chest open and my knees unbend from a defensive crouch. I stood straight again.

I had forgotten what I had always known: Food should be grounded in people and place. People telling their own stories. People seizing control of their lives and work. Growing and cooking their own food, making their own history, building their own economy. Here was a province that had known generations of inequality, and yet the people were bold and buoyant. I was proud to know them.

As the trip came to a close, I was grinning again. I was no longer angry at the sharks—some might even have a role to play. But they are not central to the story we will one day tell about the world we built out at sea. We are propelled by greater vision, more powerful than pettiness and greed.

So we headed home. I had found strength in my roots.

Harvest Time

Part of the joy of ocean polyculture is that it's always harvest season, the time when you finally get to enjoy the crops. The same species of crops carry very different flavors even a few miles apart. A blade of kelp will taste different at various times of the growing season and year to year. Fluctuating water temperatures, nutrient levels, sunlight, and other factors all interact to produce a spectrum of flavor. On the water, we call it merroir.

Kelp is robust and pretty easy to grow, but harvest time can be a bit stressful. That's because kelp hates fresh air, and as a result will begin wilting within thirty minutes of leaving the water. Our process is to hoist the kelp out of the water and sort it into three barrels with screw lids. First, we trim off the bottom tips of the blades, which are the oldest part of the plant and not so tasty. They go into the first barrel, to be turned into liquid fertilizer. Next, we move to the top of the blade and snip it just below the stipe. These food-grade blades are stuffed into the second barrel, destined either to be sold as fresh whole-leaf kelp or to be run through our noodle machine to become kelp noodles. Finally, we cut off the stipes, which are the stems of the kelp. They go into the third barrel and have a variety of culinary uses of their own.

We like to get our mussels out of the water right after the kelp harvest, in late spring, before the biofouling sets in. Getting them to shore is pretty simple: just twist off clumps and throw them into stacking totes. The problem is the beards, which are very time-consuming to get off. There are two ways to deal with this: find customers who don't mind the extra prep work—my CSF members are a good example of this demographic—or get what's called a "de-clumper." One of these can run more than five thousand dollars. Depends how flush you are after selling all that kelp!

If you're selling commercially there is a tsunami of food safety regulations. If your garden is for personal use, just don't poison your friends.

part **6**

CHAPTER 19

A Shared Vision

Coming home from Newfoundland, I felt renewed and recentered. The GreenWave team was knocking it out of the park. There were now more than thirty new seaweed farms in New England, half a dozen hatcheries, and more than two thousand acres zoned for ocean farming. Alaska farmers, only in their second year of growing, were already harvesting hundreds of thousands of pounds of kelp a year, which in turn, spurred state officials to announce a hundred-million-dollar growth plan for the mariculture industry. We partnered with a gaggle of tech wizards to embed sensors in farms around the country to track nutrient density, water temperatures, and light levels; this type of precision farming can not only increase yields, but help farmers track the positive benefits their farm is having on the surrounding ecosystem. 3Degrees, an organization specializing in carbon credits, was helping develop the first-ever framework to enable GreenWave farmers to gather and sell carbon credits. Though processing infrastructure remained a worry, several new companies were raising funds to fill the need.

But amid all this progress, a question haunted me: How were we going to avoid the pitfalls of industrial agriculture and aquaculture?

At nearly every public event, I get variations of this same basic question: How are you going to keep farmers from getting squeezed out by the Monsantos of the world? How are you going to make sure 3D ocean farming isn't converted into monoculture or used to grow GMOs? Or that leases are distributed equitably and remain low-cost? In other words, how are you going to make sure this doesn't get screwed up.

The answer is simple: I can't. But together we can. With the right mix of farmers, chefs, eaters, scientists, activists, and others, breaking new ground is possible. The good news is that, if the amount of activity on GreenWave's website alone is a measure, there are thousands upon thousands of people wanting to join us on this journey. Plus, our ocean-farming industry is still young, which grants us the luxury of historical perspective and time to build a plan before we scale. It gives us time to learn from the mistakes of the old food economy. To build lasting alliances with the fuzzy, not bloodthirsty, sharks. To work with communities to ensure that their waters are protected, not privatized. In other words, we have a chance to not fuck this up. But that's going to require a road map. To build one, we need to look to the work of others, both in and out of the seafood sector.

My first brush with goodness was Patagonia, one of the only companies operating at scale that bring to life the promise of a more just and sustainable economy. After hearing about Green-Wave's work, they reached out and asked me to spend a few days with their famed founder, Yvon Chouinard.

It was the first time I've met a no-bullshit, ethical CEO. He was gruff and rough-edged, like me. I told him I'd never set foot in a Patagonia store; he didn't care. Although he became a rich man and reached full-blown cult status among millennials, his decades of hard living hung on his bones. I felt an affinity. He told me he had been following mussel farming in Spain for decades and now wanted to jump with both feet into the restorative food world. We watched videos together of innovative farmers and fishers, using methods both old and new. Even watched some grainy handheld footage of Lummi Island reef-netting in the Pacific Northwest. When I told him for me this was "Newf porn," he burst out laughing with knowing joy. His wife, Malinda, overheard and said, "Boys, calm down."

Though we bonded about fish, it was his practical and hard-nosed vision for a new economy that made its enduring mark on GreenWave's work. This was one of my first brushes with an economic model that could work for people and the planet as a whole.

Turned out there was a whole school of thought operating as a backbone for Patagonia's ethos, which reached far beyond sustainability into the nether regions of regeneration. Vincent Stanley and John Fullerton became two of my mentors in a land known as the regenerative economy. I had been swimming with sharks; now I was sitting at the feet of lighthouse keepers.

No joke, Vincent's official title at Patagonia is "Director of Philosophy." But the job title fits the bill. His role is to inspire the company beyond the desert islands of "sustainability" and "corporate responsibility." No small feat. He was instrumental in the now famous 2011 Black Friday ad campaign. They took out a full-page ad in *The New York Times,* urging their customers not to buy mindlessly on the biggest shopping day of the year. Their "Don't Buy This Jacket" campaign highlighted actions taken by the company and challenged consumers to think before they spend. "WE make useful gear that lasts a long time," reads the ad, "YOU don't buy what you don't need."

The campaign pointed toward a different way of doing business altogether. In addition to urging customers to reduce, repair, reuse, and recycle, the ad read "TOGETHER we reimagine a world where we take only what nature can replace." These parting words suggest something beyond greenwashing—a commitment to changing the underlying logic of the economy to come into balance with natural systems.

Unsatisfied with simply reducing waste, Patagonia has moved aggressively into the food sector to catalyze the regenerative economy. Yvon now argues: "People need a new jacket every five or ten years, but they eat three times a day. If we really want to protect our planet, it starts with food." According to Birgit Cameron, managing director of Patagonia Provisions, which is the company's food arm, the next wave for the company is trying to "move beyond sustainable and embrace regenerative agriculture. In other words, sustainability connotes stasis."

Hear an echo? It's a good sign when farmers and business folks begin sounding alike.

So we started working together. It was an easy sell. Unlike so many in the food sector who avoided the stink of aquaculture, Patagonia was already bought in. They were quick to take what they were learning from land, having just finished a kernza wheat-beer project with Wes Jackson, the father of regenerative farming. They, too, have seen that this is our chance to avoid the mistakes made by industrial fishing, agriculture, and aquaculture. That we can push GMO-spiked salmon and mislabeled seafood off the table, and move bivalves and sea greens to the center of the plate. Restorative ocean farming represented Patagonia's opportunity to leverage their enlightened consumer base to ensure that those of us that work at sea can make a living and also breathe life back into the planet.

But Yvon insisted that good food grown for both people and the planet must also be delicious. One of the most enjoyable meals of my life was during an afternoon spent with a gaggle of culinary wizards at Yvon's home, tasting literally dozens of different mussel recipes. The creativity of each dish was proof that, in the right hands, mussels have the potency to be the gateway drug to a new "climate cuisine" that is both delightful and hopeful. It was like the first supper marking the future of seafood.

THE POLITICS OF YES

In 2015, I was getting pissed at the fishing industry's state of climate denial. Anyone working the ocean could see the changes: water temperature fluctuations, weirding weather, new species showing up in the nets. So I decided to travel by boat to New York for the People's Climate March. I brought along a gaggle of other fishermen and ocean farmers on the ten-hour journey, flying a flag that read "There Will Be No Jobs & No Food on a Dead Planet." Our hashtag was #climatemarchbysea.

We had a beautiful day on the water, traveling into the Narrows of New York, underneath the Verrazzano Bridge, past the floating prisons of Rikers, through Hell Gate on the East River, past the United Nations, then, finally, into New York Harbor, under the shadow of our first national immigrant—the Statue of Liberty, on a base of Stony Creek pink granite.

I had been hauling mussels the day before, so *Mookie* was properly dirty and smelled to high heaven. We pulled into the Chelsea Yacht Club next to a hundred-foot yacht. Got some sideways stares. We hopped off and hit the streets, joining more than a hundred thousand protesters.

I was expecting posters of bees and trees held aloft by liberals dressed as polar bears—the usual tone-deaf environmental antics that alienate anyone who works for a living. Not even close. These folks had backbones and answers. They came from a wide array of sectors: coal miners fighting to solarize the hollers of Kentucky, black urban farmers from Detroit growing food from a brownfield of racial injustice, nurses and First Nation communities fighting for health care and against oil pipelines. Behind these efforts stood climate scientists, unions, and enlightened social entrepreneurs.

So much on-the-ground work I could learn from. There were community land-trust activists fighting to protect rural farming land, an effort that stretched back to the civil rights movement. The first land trust in the United States was New Communities, a fifty-seven-hundred-acre reserve and farm collective in Georgia owned and operated by approximately a dozen black farm families from 1969 to 1985. From there, land trusts have grown in size and scope, with more than 250 now operating in the country. Could we create equivalent sea trusts—zones for conservation, small-scale fisheries, and restorative farming—to protect farms and prevent privatization of our seas?

All these groups used a lot of jargon to describe what they do: "just transition," "circular economy," "regenerative capitalism." To me, it was simply a solutions-based approach to change. Sure, they were fighting to stop pipelines, defend the rights of workers, and shut down coal plants that were polluting poor neighborhoods. They were using their power to play defense—to stop the bad things in the world from happening. But this new army of builders also came armed with solutions. They had their own plans to build better food systems, renewable energy systems, and community wealth.

There were urban food-justice farmers, such as the Detroit Black Community Food Security Network, an organization that runs a seven-acre farm, D-Town Farm, in northwestern Detroit. A cooperative food-buying club, the Food Warriors Youth Development Program, which aims to create youth engagement in the food system and a culture of healthy eating at local schools. Could social justice be woven into the DNA of ocean agriculture?

And co-ops were sprouting up everywhere, many funded by a group called The Working World, which provided what they call "non-extractive investment." This meant providing loans to co-ops without demanding collateral; if the business doesn't become profitable, the loan doesn't get repaid. After more than a thousand investments, supporting more than two hundred businesses, they have an extraordinary 98 percent repayment rate on these zero-collateral loans. Was this a way to fund the scaling of ocean farms?

There was also Thunder Valley Community Development Corporation, started by youth from the Pine Ridge Indian Reservation in South Dakota. They have become a model for an ecosystem approach to ending poverty. A centerpiece of their effort is a thirty-four-acre "Regenerative Community Development," which features affordable, energy-efficient housing built by Lakota construction crews, with pathways to homeownership. Gardens and greenhouses produce food that is sold at local markets. A range of art and cultural spaces support efforts such as youth development and Lakota language education. Sounds like a GreenWave Reef on land.

There were also companies out there innovating whole new ways of doing business. Organically Grown Company is a prime example of how to scale up without selling out. Over the years since their founding in 1978, they grew from a farmer-training nonprofit to a mission-driven business to a co-op to an

S corporation—not unlike our own trajectory. In the summer of 2018, they broke ground by becoming the first company to place ownership of their company in a trust. Their vice president of "organizational vitality," Natalie Reitman-White, put it like this: "Most ownership structures are still about an owner that expects to pull out profit regularly from the company as well as someday sell their stock for more than they bought it for. This stops that cycle because we'll have a single owner—a trust that will hold the company ownership stock into perpetuity—that never wants to extract a profit and never wants to sell the company." This allows founders to lock in the value of the company and not be yoked to short-term returns. Could this be the next step for GreenWave?

So many unusual bedfellows to learn from, so many questions to explore. GirlTrek was training frontline health activists to disrupt disease and inspire a new culture of physical activity in the fifty highest-need communities in America. The goal? One million black women leading a health revolution. Could we be as audacious as GirlTrek and create a million jobs in the blue-green economy?

These were folks I wanted to hang out with.

John Fullerton, head of the Capital Institute and author of the white paper "Regenerative Capitalism: How Universal Principles and Patterns Will Shape Our New Economy," a founding document of the new economy, described the power of this emerging movement in terms I could relate to: salt marshes. "There is an abundance of interdependent life in salt marshes where a river meets the ocean. At those edges the opportunities for innovation and cross-fertilization are the greatest."

In my neighborhood of Fair Haven, the edges have been described to me as the "sweetwaters," which swirl where the Quinnipiac River meets Long Island Sound. That edge, where all these people are already working, is where I want to live.

PRINCIPLES OF REGENERATIVE OCEAN AGRICULTURE

So what's our road map for the ocean? We're lucky: there are three thousand years of mistakes and innovations that we can draw from land-based farming. I have been on the water for thirty years, so I have some thoughts on where we need to go, but no definitive answers. Since the ocean will increasingly become our primary source of food, we all have a stake in the outcome and need to draft the road map together. My ten principles of regenerative ocean agriculture are designed to begin the conversation.

Circular, Not Vertical: I farm vertically, but you'd never catch me running my business that way. The sharks keep whispering in my ear that the only way to scale the industry is to build thousand-acre farms at sea and vertically integrate ownership of hatcheries, farms, processing, and product lines.

Vertical integration is at the root of a linear economy that consolidates the benefits to a few at the top. As Dan Barber, a hero of the organic food movement and co-owner of Blue Hill at Stone Barns, explains: "In the rush to industrialize farming, we've lost the understanding, implicit since the beginning of agriculture, that food is a process, a web of relationships, not an individual ingredient or commodity." We don't want one thousand-acre monoculture farm, but, rather, a hundred ten-acre farms dotting our coastline. We want diverse, circular economies producing massive volumes of food through networked production. I'm in the business of building reefs, not skyscrapers. A "GreenWave Reef" is made up of fifty ocean farms, clustered around a seafood hub and hatchery, surrounded by a ring of institutional buyers and entrepreneurs. Unlike a vertical economy, the reef model distributes rather than concentrates profits, encourages the sharing of infrastructure, and helps keep capital and ownership local.

Who Farms Matters: As environmentalists, we want to see more kelp and mussels in the water to rebuild reefs and soak up carbon. As eaters, we want more of these and other restorative crops on our dinner plates, because they are heart-healthy and delicious. But in this era of extreme economic inequality, it matters *who* is farming the ocean. If we leave it up to business-as-usual, it'll be large corporations that monopolize the industry and reap the benefits, leaving scraps for the rest of us.

If we're going to build a new economy, then who farms matters. I started GreenWave to make sure poor and working people are on the front lines of the blue-green revolution. These groups range from farmers and fishermen pushed off the land and sea by globalization and climate change to First Nations and other marginalized groups who have been systematically excluded from sharing in the benefits of the industrial revolution. I don't care if you're a former felon or an undocumented immigrant—as long as you work hard, you have the right to a good, meaningful job.

For the majority of our country, there's no ladder of economic justice. We have to build it for ourselves. This means things like keeping ocean leases affordable, to ensure that people from all walks of life can start farms; limiting the acreage any one entity can own, to undermine monopoly power; and mandating living wages and open employment on farms and processing plants.

Protect Ocean Commons: I am not a conservationist; I believe firmly in farming our seas. But our oceans are beautiful, pristine spaces owned by society as a whole—and we need to keep them that way while still opening channels for ocean farmers from all walks of life. To achieve this balance, we need, just like community land trusts, a network of sea trusts in every coastal state. These sea trusts will include conservation zones, networks of ocean farms, and areas dedicated to sustainable wild fisheries. The farms will breathe life back into both the wild fisheries and the marine parks. And sea trusts will not be privatized spaces but, rather, places where anyone can boat, fish, or swim, and communities will retain a lever of community control through regular lease renewal. This will allow sea trusts to ensure protection of our local waters, while empowering a new generation of ocean farmers and fishers to preserve the health and welfare of marine species.

Value Ecosystem Services: While others pollute, we restore—and as farmers we should be paid for the positive externalities of our work. In the new economy, markets have to reflect the environmental value of restoration. Thanks to the work of The Algae Foundation, in February 2018, the Senate included a carbon tax for macroalgae for seaweed in the farming bill. It was too low—thirty-five dollars per ton—but a historic victory for recognizing the ecosystem services provided by ocean farming. A similar effort is under way at the state level to open nitrogen-trading programs for shellfish farmers.

We need to support these efforts, because they will create incentives to farm in urban waters. This type of "pollution farming" doesn't grow food, but, rather, harvests ecosystem services, including the removal of nitrogen and carbon as well as heavy metals. These pollution farms have already been built in the Bronx River and the Port of San Diego. More are to come.

Adopt Appropriate Technology: We all know the robots are coming. Over the years, I've become increasingly suspicious of technology—I abandoned Twitter and Facebook long ago. But I've begun to soften a bit, and now believe we should kill some, but not all, of the robots.

As I write, there is a gaggle of U.S. engineers designing autonomous ocean-farming boats and harvesters. Sorry, but I'm gonna drive my own fucking boat. As soon as one of those are put in the water, I'll be sneaking out with my shotgun on a foggy night to put a hole in its hull—a quick, painless death.

But some robots we're going to need. The central challenge of ocean farming is that our "soil" turns over a thousand times a day, and we can't see what we grow. This means we need embedded sensors in our farms and have to deploy autonomous underwater "fish" to keep an eye on our crops. We need technology to track underwater weather. We need real-time data uploaded to online dashboards so we can know precisely when to plant, where in the water column the nutrients are the densest year to year, and how to keep pollutants from showing up in our crops. We need mechanical fish alerting us to early biofouling to optimize harvest times. These cutting-edge tools of predictive farming will help us increase yields, monitor climate change, and speed up the learning loops as our industry scales up.

In the process, this technology will allow us to harvest more than crops—we can harvest data, to be sold to scientists monitoring acidification, companies selling carbon and nitrogen credit, and wholesale buyers needing to predict supply yields. Data harvesting will translate into an entirely new income stream for farmers.

Balance the Scales: For too long, farmers and fishermen have been caught in the beggar's game of selling raw commodities while others soak up the profits.

We need to rearrange the relationships between those of us

who produce food and those of us who buy it. Failure means re-creating the power dynamics of the old economy. This new relationship requires leveraging new business models, such as purpose trusts, and organizing nationwide digital co-ops so farmers can negotiate equitable prices for their crops on the docks. It also means forward contracts, allowing farmers to get paid before they grow; if our crops fail, then both the farmer and the buyer share the loss. And financial institutions need to begin providing zero-collateral loans for farmers to scale. It's time for everyone to shoulder the risk in the risky business of growing food in the era of climate change and globalization.

To achieve the dual goals of protecting people and the planet, the blue-green economy cannot be underwritten by "finance-as-usual." It's time for those with capital to divest from the old economy and invest in the new. This will mean less profit for the private sector, but it will also mean more value in terms of social and environmental good.

Radical Traceability: It makes me heartsick that our nation's food is smothered in lies. Inventories and sales of illegal and mislabeled seafood are at an all-time high throughout the entire domestic marketplace; one major recent study revealed that nearly 50 percent of seafood subjected to DNA testing was something other than what the labels or menus claimed. And even the best-known sustainable brands are neck-deep in fraud.

This has to end, and, unlike most problems we face, learning where your food comes from is a cheap, easy, and vital hack. Dock to Dish has spearheaded the traceability revolution by pioneering the world's first live tracking dashboard to monitor hauls of seafood, with precision and accuracy, from individual fishermen and growers at sea directly to end consumers on land, in near-real-time calibration. Right out of the gate, we need to make traceability an unbendable principle by applying these groundbreaking technologies to our crops.

Beyond Organics: Although you'll sometimes see organic labeling on some sea greens and shellfish products, the whole concept, when applied out to sea, is a nonstarter. With tides, currents, and storms, the ocean is too dynamic a growing environment for us to import the land-based framework of organic certification. Soil is stable; inputs are predictable. But imagine if soil moved freely from farm to farm—organic standards would be fruitless. As with so many aspects of ocean agriculture, the seafood industry has lazily adopted land-based standards that are a poor fit out at sea.

The oceans require a unique framework, which, like Patagonia, I'd classify as regenerative. Obviously, this means cutting out GMOs, chemicals, and herbicides, but, most important, regenerative certification needs to mandate stringent water-quality testing, seed-to-harvest traceability, labor standards, and crop selection. While the standard of sustainability may apply to fish, regenerative certification should be limited to crops that restore ecosystems, such as seaweeds and shellfish.

Collaboration, Not Competition: There is a co-op of seaweed farmers in Japan that meets every December to share lessons learned. Each year, members are assigned one problem to solve. They spend the year experimenting and documenting, and return in December to report their progress. They do the same thing year after year. I love this example, because it's founded on the principle that collaboration, not competition, spurs innovation and adoption of new practices. It is understood that rapid cycles of learning make for a nimble industry that internalizes the best practices and greases the wheels for scale.

Grow with the Ocean: The marketplace should not dictate what farmers grow—it should be the oceans that decide. One of the significant missteps of traditional aquaculture has been growing what people want to eat, not what the ocean can provide. As we scale up, we must use polyculture, not monoculture, farming methods and grow ecologically appropriate species. With climate change heating up our waters and driving species ever farther northward, "ecologically appropriate" will not always mean "native," which assumes nature exists in a static state. It doesn't. And there are plenty of native species that are growing increasingly destructive—such as Pacific urchins—and should not be farmed. Instead, in collaboration with environmentalists and scientists, we need to be constantly evaluating what is healthiest for the ecosystem to grow.

CHAPTER 20

Fade to Blue

My travels through the food movement have convinced me that food has a role in saving us as a nation. Food is the largest private employment sector in the country, home cooking is in vogue, and the tent of the food movement continues to expand. Regular gas station eaters like me are feeling a deep affinity with the vision of brilliant food leaders like René Redzepi and Mark Bittman. Even more encouraging are the generations of youth hell-bent on changing the food system from below.

I hope you sense from this book that there is an alternative route, both for our oceans and for our economy as a whole. We can farm to restore, not deplete, while creating soul-filling jobs for millions of people, breathing life back into our oceans, and exploring the uncharted world of underwater tastes. No matter what the captains of industry tell us, there is hope.

These aren't mere pipe dreams. The hard work of others is quickly turning vision into reality. I started GreenWave to take myself out of the equation—my own version of planned obsolescence. I am not the agent of change; my work is done. Farmers like Jay and Suzie, Jill, Dave, and Tomi are the new faces of the blue revolution. New leaders, such as Emily, Kendall, Karen, and Liana, will take ocean farming to a new level. My dream is to visit a GreenWave farm ten years from now and marvel at the vertical layers of innovation.

GreenWave has big plans: five hundred farms in five years. That means ten regional reefs, each composed of fifty farms. We're already supporting new reefs in California, New York, the Pacific Northwest, and Alaska, even Europe. We're wrapping up an open-source farming toolbox, bringing together farmer-to-farmer "chat and chews" to swap tricks of the trade, and organizing a national digital co-op so buyers can't drive down prices by pitting farmers against one another. Urban ocean pollution farming is now a reality in San Diego and New York, as well as a framework for assigning a value to the carbon sequestered by kelp. What's emerging is a farm of the future that harvests not just food, but also data to sell to scientists and ecosystem services to sell as credits.

And markets are expanding quickly. Patagonia is rolling out a new line of restorative ocean foods, starting with canned mussels and expanding to kelp bouillon cubes. Ocean Rainforest, Google, Ocean Approved, and Blue Evolution are major buyers of farmers' crops. Companies like Akua are churning out kelp-and-mushroom jerky, and Stony Brook University and Elm Innovations are wrapping up product trials on seaweed-based fertilizer and cattle feed.

To pull off this grand vision, we need all hands on deck. Here are a few ways to get involved:

Organize a house party: Plan a local event for friends and family to learn about the ecological benefits and culinary potential of ocean farming. On the GreenWave website, there is a link to a house party resource page where you can find videos to show your guests, talking points, and ways to order kelp for the dinner.

Get cooking: The recipes at the back of this book will give you a head start, but we need you to start cooking with sea greens. Get creative, invent new dishes that desushify seaweeds. Know that every meal you cook increases demand for our crops. To buy sea greens to cook with, visit Blue Evolution's website to order directly from their online store. Also, check out brands like Akua, which makes kelp jerky; Ocean Approved and Sea Tangle for seaweed pastas; and Barnacle Foods for its kelp salsa and pickles. Many of these new products are available online and at Whole Foods, Trader Joe's, and local natural and specialty markets. Dried seaweeds are widely available at most grocery stores. Keep an eye out for Patagonia's new line of seaweed spices and umami-bomb bouillon cubes.

Use seaweed and shellfish to fertilize your garden: Even the best organic fertilizers are dumping nitrogen into our waterways, to end up in the ocean. Try using seaweed fertilizers to lessen the impact of your garden. Your plants will love it.

Help revise the Principles of Regenerative Ocean Agriculture: Treat the principles in Chapter 19 as a living document to be continually torn apart and revised. It's just a draft. Help us make it better.

Create local Sea Trusts: Work with your town and state agencies to organize dedicated zones for restorative ocean farming so farmers have pre-selected areas to grow in that have been vetted by the community. Fight to keep lease rates low, so the doors stay open to beginning farmers and large-scale corporations can't stockpile leases.

Support NOAA: NOAA gave birth to domestic shellfish farming in the 1930s and continues to fund innovations such as seaweed farming, thereby opening up new horizons for unemployed fishermen and their children. Same with its Sea Grant programs, which provide education and technical assistance for farmers and ocean-based industries. Supporting these agencies come federal budget time translates directly into supporting farmers like me.

Start your own hatchery, farm, or underwater garden: Get your feet wet by building a small hatchery or building your farm locally. GreenWave is developing lots of online resources. Check out the University of Connecticut and Ocean Approved manuals for building a DIY kelp hatchery.

Get a degree in ocean agriculture: If you're heading off to college or looking for a new career, make ocean agriculture your focus. The Algae Foundation has developed a degree program for community colleges in macroalgae, and graduated its first class in 2018. For those interested in advanced degrees, places like the University of Connecticut and the University of California, Davis, are doing cutting-edge research in everything from kelp-raised beef and selective breeding to ocean policy and economic development.

Support GreenWave: A simple and direct way to help is to support GreenWave's work. The team is building ten reefs, consisting of five hundred farms, in the next five years. They can't do it alone. So donate, volunteer, sponsor a farmer—whatever you can do to help.

HAIL AND FAREWELL

On the wall of my office hangs a wood-framed picture taken in 1898 of Newfoundland sealers. These are hated men, made criminal by animal rights activists. I bought it at a junk shop on Duckworth Street in St. John's after the cod stocks crashed, just as thousands of fishermen were being thrown out of work.

The photo shows twenty men. Hats, beards, mustaches, vests, suspenders, wool pants, and tweed coats. Pride in their eyes. A barrel-chested captain leans on the mast. The cook, in his bloody apron, stands next to the first mate. In the front row sit two brothers holding hands. The younger, clearly under his brother's protection, has stooped and fragile shoulders. His right hand lies limp in his brother's huge paw. When I brought it to the cash register, the junk shop owner, steeped in Newfoundland history, pointed to the younger brother and said, "That one died on the trip."

These men lived good lives. No matter how brutal the work, they refused to leave the water. I am one of them. My meaning comes from the sea. Yes, my hunting days are over, but as an ocean farmer I am still graced with salt on my lips and skeleton shrimp in my beard. I am happiest breaking ice off the deck of *Mookie*, checking kelp lines and scallop nets when the wind and waves are beginning to kick up. I have the heart of a fisherman and the soul of a farmer.

Soon it will also be my time to fade to blue. I can hope for no better than to grow old tending my small patch of water, and one day to drift quietly and willingly into its embrace. And if the fishermen before me, taken by the sea, give an approving nod to my journey, I too will have lived a good life.

Ocean Green Profiles

～～～～～～～～～～～～～

SUGAR KELP

Sugar kelp has a single, elongated blade and a long, thick stipe. The blade is light brown in color, with a wave of undulating frills along its margins. It is sweet, mildly salty, and packed with umami, reminiscent of an early green bean. The blade retains a pleasant crunch even when cooked, and the bittersweet stipe lends itself perfectly to pickling. When it hits boiling water, sugar kelp transforms instantly into a startling bright green. Gently blanched, it can be slivered into hot or cold pastas, and also makes a savory addition to deviled eggs or quiche, or a salty bite in a green salad. It is excellent cooked with vegetables or beans, bringing out a brightness and depth of flavor without changing the taste. This magical ability to highlight natural flavors comes from the abundant glutamate that is the basis of umami. In fact, kelps have the highest concentration of glutamate of any plant on the planet.

OARWEED

Oarweed takes its name from its broad, oar-shaped stem. It clings to rocky substrates with a clawlike holdfast, and ends in a cluster of smooth blades. These blades are packed with glutamate, lending them a savory marine flavor with smoky notes. Different preparations transform oarweed's texture and appearance: raw, it is thick and fleshy and dark brown; quick heat turns the blades a vibrant shade of green; boiling renders them translucent and tender. Oarweed is one of the best species for drying and grinding into flours to make quick breads, scones, cakes, and cookies.

WINGED KELP

Winged kelp's name comes from the wide, spore-bearing leaves that flank the edible blade. It is also known as "sea mustard" and "Atlantic wakame." The blades range in color from olive to tan, with a lighter stem giving structure to the otherwise ruffled and membranous leaf. Considered the most tender of sea vegetables, it turns green when cooked, and has a spinachlike flavor with nutty overtones. When reconstituted from its dried version, it has a rich texture and expands to seven times its original size. After it has been soaked, then cooked, its texture turns silky, and it almost melts in your mouth. It can substitute for spinach in any preparation, from spanakopita to curry, and can also be grilled, toasted, or mixed into soups.

BLADDERWRACK

Bladderwrack is one of the better-known ocean greens. Air-filled bladders stud its blades, ranging from pea-sized to marble-sized, and increasing in number with the turbulence of the waves. Between bladders, the fronds are flat and smooth, and range in color from spring green to a light chestnut. The frond has a strong flavor reminiscent of salty fish, and is best as the starring character of a dish. It's excellent in broths and sauces as a counterbalance to acidic flavors and fats.

GRACEFUL REDWEED

Graceful redweed grows in clumps of beautiful, dark red branches, and creates a tangle of smooth, spaghetti-like strands underwater. It is an efficient producer of agar-agar, a translucent, neutral-tasting substance that is often used as a thickening base for sweets, jams, pies, and puddings. In the kitchen, redweed is best prepared fresh and cold, to bring out its delicate flavor and its celery-like crunch.

IRISH MOSS

Irish moss has branching fronds whose color alters according to its environment, from scarlet to greenish yellow. Irish moss has largely been cultivated for its yield of thickening carrageenan, which is used in everything from ice cream to lunch meat. However, with its sweet, intense shellfish flavor, it is delicious on its own. It can be used raw, cooked, or pickled, and is a great accompaniment to fish or pasta.

DULSE

Clinging to rocks or other kelp species with small holdfasts, dulse produces furrowed, magenta fronds. When it is dried to a leathery texture, dulse is often referred to as vegan bacon or jerky for its satisfyingly chewy texture and rich flavor. Fried or toasted, the leaves become remarkably crisp and delicate. Cooked, it takes on a wonderful buttery texture. Dulse can also be eaten raw, ground into flakes for a savory topping for rice or popcorn, or added to soups and stews. Dulse is an especially glutamate-rich seaweed, and can boost the flavors of meats and vegetables. It's a workhorse ally to many dishes, as well as a satisfying snack on its own, and its versatility has made it a popular ocean green from Scandinavia to the U.S. Eastern Seaboard.

LAVER

Laver sprouts a number of sheer reddish-purple fronds from a central holdfast. These fronds are typically minced and laid out to dry to create the classic sushi "sheet" form, taking advantage of laver's natural thinness. Though laver is most familiar as the seaweed encasing sushi rolls, it has a long tradition of diverse culinary uses. It adds a briny-sweet olive note to a pâté, holds up well to baking or smoking, and melts into batter for frying fish or vegetables. Toasting brings out a sweet and nutty flavor, and imparts a green tint to the purple fronds. In Wales, laver is commonly eaten for breakfast with bacon and cockles, and for dinner with lamb and monkfish. As a result, it's sometimes called the "Welshman's caviar."

SEA LETTUCE

Sea lettuce is akin to common land lettuces in form and function: delicate, frilly, and best enjoyed fresh, because cooking can turn it bitter. The first bite brings a briny burst to the tongue, which mellows to a bright, lemony flavor. Though being eaten raw highlights sea lettuce's best properties, cooks have also added it dried to compound butters or finishing salts, where it can impart hints of truffle and vanilla.

SEA BEANS

Sea beans' succulent branches, broken into pieces, bear a passing resemblance to their namesake, the green bean. Steamed and coated in olive oil, they make a single-ingredient salad. Sea beans can be readily substituted for asparagus or green beans, and, as with these vegetables, care should be taken to avoid overcooking in order to maintain their crispness and bright color. They can bring a strong, deeply salty flavor to dishes.

SEA PALM

Sea palm, also known as American arame, resembles a miniature palm tree with dark caramel strands. The sweetest and subtlest of all the ocean greens, prized for its versatility, it quickly softens and expands when cooked. It has a widespread array of uses, and can be added to everything from pilafs to trail mix. Though prized for its mildness, it can hold its own as a side to heavier meat dishes. Like many of the more delicate ocean greens, it responds beautifully to an overnight soak in a favorite marinade.

SARGASSUM

Sargassum ranges from dark green to walnut brown and has a bushy shape with narrow blades. Its nutty flavors are very subtle and, as a result, highly versatile. In Hawaiian cuisine, sargassum is served fresh with octopus, baked in fish casseroles, or rolled into fillets of fish and roasted. It can also be mixed with brown sugar and used as a filling in steamed buns, or combined with oil, salt, and green onions as a filling for dumplings. Deep-fried in tempura batter, sargassum turns into a delicate chip.

Recipes

To buy sea greens for cooking, you can order directly from Blue Evolution's website at www.blueevolution.com.

Brooks Headley Recipes

BARBECUE KELP AND CARROTS

¼ cup olive oil
1 white onion, finely diced
2 cloves garlic, thinly sliced
2 cups water
½ cup maple syrup
½ cup tomato paste
½ cup white wine vinegar
2 teaspoons molasses
2 tablespoons yellow mustard
1 tablespoon hot sauce (your choice)
Salt
Freshly ground black pepper
Smoked paprika

· Sweat the onion and garlic with ¼ cup of olive oil until soft.
· Add the rest of the above ingredients, and simmer slowly until the sauce is reduced and thickened.

    1 pound large carrots, sliced on the diagonal
    ¼ cup olive oil
    Salt and pepper, to taste
    ½ cup fresh kelp ribbons
    ½ cup labneh or Greek yogurt, fresh
    Crushed corn chips
    ½ pound mustard greens
    Freshly squeezed lemon juice
    Bread crumbs, as garnish

· Roast carrots at 400 degrees Fahrenheit with the olive oil
  and salt and pepper until softened and charred around
  the edges. Toss together with the kelp and barbecue sauce
  until coated. Top mixture with a dollop of yogurt and some
  crushed corn chips. Dress mustard greens lightly with a
  good squeeze of lemon juice and salt, and add on top of
  the carrot-and-kelp mixture. Sprinkle bread crumbs over
  the top, and dress with a squirt of olive oil.
· Serve immediately.

KELP NOODLES WITH TAMARIND

AND PEANUTS AND SEARED TOFU

This is our version of take-out pad thai, with the kelp "noodles"
taking the place of the traditional rice noodles.

    3 tablespoons tamarind concentrate
    2 tablespoons rice vinegar
    ½ cup soy sauce
    ½ cup water
    ¼ cup dark-brown sugar
    2 tablespoons sriracha

· Combine all the above ingredients, and bring to a boil. Allow to cool.

    2 cloves garlic, sliced
    1 tablespoon peanut oil, plus more for cooking tofu
    1 cup sliced red bell pepper
    16 ounces firm tofu, cut into ½-inch dice
    16 ounces kelp noodles
    1 bunch scallions, chopped
    1 red onion, thinly sliced, soaked in lemon juice for 1 hour
    ½ cup bean sprouts
    Cilantro
    Juice of 2 limes
    Chopped roasted peanuts

· In a large sauté pan over high heat, cook the garlic in a tablespoon of peanut oil until lightly browned. Add the sliced bell pepper, and cook until slightly charred, just a few minutes.
· Decant onto a sheet pan, and allow to cool.
· To the same hot pan, add the tofu and a bit more peanut oil, and cook until golden and crisp.
· Add the red-pepper-and-garlic mixture back into the pan, and toss to combine.
· Add kelp noodles, and toss to combine.
· Add ½ cup of tamarind sauce off heat, and toss until fully coated.
· Top with scallions, some of the drained sliced red onion, the bean sprouts, a handful of cilantro, the lime juice, and a dusting of the roasted peanuts.
· Serve immediately.

ZUCCHINI AND KELP CAKE WITH SEAWEED AIOLI

This is a vegan variation on crab cakes, using kelp as the sea element, with a creamy eggless mayo studded with seaweed.

> 3 tablespoons golden flax seed, ground
> ½ cup water

- In a medium saucepan, combine the ground flax and water, and heat, stirring occasionally, until you have a thick slurry.

> 1 tablespoon Dijon mustard
> 1 tablespoon cane sugar
> 2 teaspoons salt
> ¼ cup cider vinegar
> 2 cups grapeseed oil
> ½ cup chopped kelp

- To the bowl of a food processor with the blade attachment, add the flax slurry along with the mustard, sugar, salt, and vinegar.
- Pulse and stream in the grapeseed oil until a stable mayolike consistency is achieved.
- Add the kelp, and process until it's studded throughout the dressing. The kelp will not totally break down, but will remain in tiny pieces in the mayo. This makes more than you will need, but it holds well in the refrigerator for 1 week.

> 2 cups grated zucchini
> 1 teaspoon salt
> 1 cup bread crumbs
> 2 teaspoons Old Bay Seasoning
> 2 tablespoons seaweed aioli (see above)

1 teaspoon Dijon mustard

Freshly ground black pepper

3 tablespoons chopped fresh parsley

1 cup chopped kelp

A few glugs extra-virgin olive oil, for sautéeing

Fresh lemon juice, to garnish

· Combine the grated zucchini with the salt, and set in a strainer; place a ziplocked bag filled with ice on top of it.
· Allow the zucchini to drain for 30 minutes.
· Mix the drained zucchini with the remainder of the ingredients (except the lemon juice) until combined and sticky.
· Form into patties, and sauté in olive oil over high heat until golden on both sides.
· Top each cake with a dollop of the seaweed aioli and a squirt of lemon juice.
· Serve immediately.

BROWN RICE SALAD WITH TAHINI YOGURT, FRIED KELP, AND POMEGRANATE JALAPEÑO SYRUP

A fresh-tasting, nutty rice salad with Israeli flavors and fried bacony-tasting kelp as the condiment.

2 cups brown rice

Salt

3½ cups water

· Toast the brown rice over medium heat until nutty.
· Add the water and salt, bring to a boil, and then drop the heat to a simmer. Cook until all the water is gone.

· Remove from heat and cover, allow to steam off heat for 30 minutes.

1 cup tahini
Juice of 4 lemons
Water as needed
2 tablespoons maple syrup
½ teaspoon toasted sesame oil
Salt
Freshly ground black pepper

· Combine all ingredients in a food processor fitted with the blade attachment, thinning and emulsifying with water as necessary to obtain a smooth, silky purée.

3 cups grapeseed oil
1 cup kelp noodles
Kosher salt

· Heat oil to 375 degrees in a tall, heavy pot.
· Pat kelp dry between paper towels. Fry one piece at a time. Be careful—there will be oil spatters.
· Drain on a rack, and season each round with salt. Reserve.

1 cup pomegranate molasses
3 tablespoons roasted chopped jalapeño

· Combine, and allow to sit. The jalapeño will release water and thin out the molasses.

½ cup chopped fresh scallions, to garnish

· Mix the tahini dressing with the brown rice, and toss to combine. The mixture should be dressed but not soupy.
· Top with the fried kelp.
· Drizzle the mixture with the jalapeño syrup, and top with the chopped scallions.
· Serve immediately.

KELP AND CAULIFLOWER SCAMPI

Shrimp scampi is everyone's favorite Italian American dish. Here we replace the shellfish with kelp "noodles" and bulk it up with a good-for-you brassica, in the form of cauliflower. The sauce is vegan and contains no butter.

5 cloves garlic, thinly sliced
2 cups dry white wine
¼ cup freshly squeezed lemon juice
2 cups extra-virgin olive oil
1 pinch carrageenan or xanthan gum
Chili flakes, to taste
Fresh parsley, to taste, plus more to garnish
Salt, to taste
Freshly ground black pepper, to taste
Capers, to taste
2 pickled cherry peppers, with 2 tablespoons brine

· Cook the garlic in the white wine, and boil off the alcohol.
· Add lemon juice and olive oil, remove from heat, and emulsify with a whisk.
· Add xanthan gum, and whisk until slightly thickened, to the consistency of warm butter.

- Season with chili flakes, fresh chopped parsley, salt, and pepper.
- Dice a few capers and the cherry peppers, and add to taste.
- Finish with the pickle brine. This will make more sauce than you need, but it holds well in the refrigerator for a week.

  1 tablespoon extra-virgin olive oil
  1 pound cauliflower, cut into florets
  ½ pound fresh kelp

- In a large sauté pan over very high heat, heat the olive oil and the cauliflower in one layer. When the cauliflower has softened slightly and browned, add kelp "noodles" and toss to combine.
- Remove from heat, and add 1 cup of the scampi sauce, so that the kelp and cauliflower are well coated. Add chopped fresh parsley, and serve immediately.

Dave Santos Recipes

SHRIMP FRA DIAVOLO WITH KELP

  2 tablespoons extra-virgin olive oil
  1 cup finely diced onion (from 1 medium onion)
  5 cloves garlic, sliced
  1 tablespoon chili flakes
  One 16-ounce can crushed tomatoes
  1 teaspoon sugar
  ½ tablespoon salt
  4 cups kelp noodles
  12 large raw shrimp, shelled, deveined, and diced

12 leaves basil, torn
Grated Parmesan (optional)

· To a medium saucepan over medium heat, add the olive
oil and onion. Sauté the onion for about 3 to 5 minutes,
until softened and slightly colored. Add the garlic and chili
flakes, and sauté for another minute. Add the tomatoes,
sugar, and salt, and bring to a boil. Once it's boiling, lower
the heat to a slow simmer, and cook for 15 minutes. Taste
for seasoning at this point, and adjust as needed. Add the
kelp, and heat through, about 5 minutes (or to desired
tenderness), and then add the shrimp. Cook gently for
another 4 minutes, until the shrimp are cooked and tender.
Fold in the basil, and serve. Put fresh Parmesan on top if
desired.

KELP BUTTER

2 tablespoons extra-virgin olive oil
1 cup diced onion (from 1 medium onion)
10 garlic cloves, thinly sliced
Sea salt
Freshly ground black pepper
2 cups chopped sugar kelp
1 pound high-quality butter, at room temperature

· Pour the olive oil into a sauté pan, and place the pan
over medium heat. Add onion and garlic, season with
salt and pepper to taste, and sweat the vegetables until
they're soft and transparent. Add the kelp, and sauté
until it changes from dark green to lighter greenish brown

and melds together. Allow to cool slightly, then add to a food processor along with the butter, and blend until well incorporated. Season with salt and pepper to taste, and serve with crostini or warm bread.

KELP AND ORZO SOUP

4 tablespoons extra-virgin olive oil
1 cup diced carrots (from 2 large carrots)
1 cup diced onion (from 1 medium onion)
1 cup diced celery (from 4 stalks celery)
3 cloves garlic, chopped
2 cups chopped kelp
1 tablespoon salt
1 teaspoon freshly ground pepper
2 quarts vegetable stock
1 cup orzo

· In a large stock pot, over medium heat, heat the olive oil. Add the carrots, onion, celery, and garlic. Sweat the vegetables for about 10 minutes, until they have softened. Add the kelp, salt, pepper, and vegetable stock. Bring to a simmer, and simmer gently for 30 minutes, until all the flavors have melded together. Add the orzo at this point, and simmer until it's tender, about 10 minutes. If soup gets a little thick, you can add a little more stock to adjust.

Acknowledgments

I'll never be certain why Kim Witherspoon, my agent, and Lexy Bloom, my editor at Knopf, took such a big risk on me, but it was an incredible honor to work with them both. Deep thanks to Peter Meehan for being the catalyst for this book.

Thanks to Emily Stengel and the GreenWave team for bringing a new energy and vision to the work, and to Charlie Yarish for being a friend and a mentor. I am also deeply indebted to the work of NOAA, the national Sea Grant system, and all the scientists and ocean innovators who were working long before I showed up on the water. And to the team at the Woods Hole Oceanographic Institute, including Scott Lindell and Dave Bailey.

Special thanks to Lisa Fitch, owner of the Quinnipiac River Marina and one of my ocean heroes, for opening her doors to GreenWave. To the folks on the docks who were instrumental in getting me through the lean years, I hold a tally of favors and debts. The legal wizards at Gibson Dunn, especially Kevin Kelly and Dylan Cassidy, for fighting off the sharks. Lisa Holmes and Rick Meyer for their early support. Nathalie Laidler-Kylander and Karen Simons for their mentorship. Rachel Plattus, Brendan Bashin-Sullivan, and Gracie White for their research assistance. And Jessica Purcell, Tom Pold, and Sara Eagle from Knopf.

Thanks to Mark Bittman, Paul Greenberg, and Thomas Harttung for their leadership in this movement, and Vincent Stanley at Patagonia, Kenny Ausubel at Bioneers, Schmidt Family Foundation, and John Fullerton at Capital Institute for guiding GreenWave's vision. To Sean Barrett at Dock to Dish for bringing passion and storytelling skills to the work.

So many organizations have supported my work, including Echoing Green, Ashoka, Buckminster Fuller Institute, Bioneers, LIFT Economy, New Economy Coalition, Claneil Foundation, Draper Richards Kaplan Foundation, Northwest Atlantic Marine Alliance, and Yale Sustainable Food Program. I came into so much of this work knowing next to nothing, and this network of support has kept me afloat.

On the culinary front, my gratitude to chefs like Jason Sobocinski, René Redzepi, Dave Santos, Barton Seaver, Brooks Headley, Avi Szapiro, and Galen McCleary at Patagonia Provisions, for bringing their creativity to the challenge of making kelp the new kale.

This book springs from my mom's love above all. Every day on the boat, I miss her. Thank you to Sylvia, my stepmom, who has been a force in my life for thirty years. Thanks to my extended family for putting up with me. Finally, to my wife, Tamanna, my ruthless reader and true love.

Notes

INTRODUCTION

9   "I've met many female fishermen": Ilima Loomis, "'Fishers' or 'Fishermen'—Which Is Right?," *Hakai Magazine,* Oct. 13, 2015, https://www.hakaimagazine.com.

10 "sequoia of the sea": Dana Goodyear, "A New Leaf," *The New Yorker,* Nov. 2, 2005.

11 According to the World Bank: Rasmus Bjerregaard et al., "Seaweed Aquaculture for Food Security, Income Generation and Environmental Health in Tropical Developing Countries (English)," The World Bank, July 1, 2016, http://documents.worldbank.org.

11 Totaling the size of Washington: Roelof Kleis, "Growing Seaweed Can Solve Acidification," Phys.org, Dec. 23, 2010, https://phys.org.

11 And farming 9 percent: Tim Flannery, *Sunlight and Seaweed: An Argument for How to Feed, Power and Clean Up the World* (Melbourne: Text Publishing, 2017), p. 141.

11 "We must plant the sea": Jacques Cousteau, interview, July 17, 1971, quoted in Robin Neill, "Aquaculture Property Rights in Canada," in *A Breath of Fresh Air: The State of Environmental Policy in Canada,* ed. Nicholas Schneider (Vancouver, B.C.: Fraser Institute, 2008), p. 180.

12 To keep pace with rising population: Ned Potter, "Can We Grow More Food in 50 Years Than in All of History?," ABC News, Oct. 5, 2009, https://abcnews.go.com.

12 agriculture already uses 70 percent: Tariq Khokar, "Chart: Globally, 70% of Freshwater Is Used for Agriculture," The World Bank, March 22, 2017, https://blogs.worldbank.org.

12 More and more of our food system: Thin Lei Win, "Industrial Fishing Happening over Half of the Oceans, Say Scientists," *Reuters,* Feb. 23, 2018.

12 In Asian waters alone: Boris Worm et al., "Impacts of Biodiversity Loss on Ocean Ecosystem Services," *Science* 314 (2006), pp. 787–90.

12 "The sea, the great unifier": Bridget Nicholls, "How Cousteau Inspired My Love of the Oceans," BBC News, Nov. 20, 2010, https:// www.bbc.com.

13 "All of us have": John F. Kennedy, "Remarks at the America's Cup Dinner Given by the Australian Ambassador, September 14, 1962," John F. Kennedy Presidential Library and Museum, Sept. 14, 1962, https://www.jfklibrary.org.

13 Every other breath we breathe: Mónika Naranjo González, "Sea to Space Particle Investigation: Every Other Breath," Schmidt Ocean Institute, Feb. 9, 2017, https://schmidtocean.org.

CHAPTER 2: SALTWATER COWBOYS

33 "So many commercial fishermen": Mario Vittone, "Letters: 'Reply All: The 1.5.14 Issue,'" *New York Times Magazine,* Jan. 17, 2014.

CHAPTER 6: EVERYONE'S ON THE HOOK

53 When I was on the Bering Sea: Mark Kurlansky, *Cod: A Biography of the Fish That Changed the World* (New York: Penguin, 1998), p. 117.

53 "It's no fish ye're buying": Ibid., p. 175.

54 fishermen aren't in control of fish: Seth Macinko and Brett Tolley, "Fish and Ocean Grabbing: The Case of Commercial Fisheries," slowfood.com, April 6, 2014, http://slowfood.com.

54 In 2018, another Wall Street: Beth Brogan, "Fleet of 5 Maine Fishing Trawlers Sold to New York–Based Equity Firm," *Bangor Daily News,* Dec. 4, 2018.

55 There the entangled catch is devoured: Mission Blue, Sylvia Earle Alliance, "Ghost Nets, Among the Greatest Killers in Our Oceans . . . ," mission-blue.org, May 13, 2013, https://mission -blue.org.

55 We own more water rights: Paul Greenberg, "Why Are We Importing Our Own Fish?," *New York Times,* June 20, 2014.

55 "California squid are being caught": Paul Greenberg, "The Long Journey of 'Local' Seafood to Your Plate," *Los Angeles Times,* July 11, 2014.

56 a side glance by the FDA: Food & Water Watch, "Toxic Buffet: How the TPP Trades Away Seafood Safety," Foodandwaterwatch.org, Oct. 2016, https://www.foodandwaterwatch.org/sites/default/files/rpt_1609_tpp-fish-web_2.pdf.

56 one in three fish: Food and Agriculture Organization of the United States, "The State of World Fisheries and Aquaculture 2018—Meeting the Sustainable Development Goals," fao.org, Dec. 2018, http://www.fao.org/3/I9540EN/i9540en.pdf.

56 According to the chef and author Mark Bittman: Andy Sharpless and Suzannah Evans, *The Perfect Protein* (New York: Rodale, 2013), p. 41.

CHAPTER 8: SEA OF SUNKEN DREAMS

68 between 2000 and 1000 B.C.: Fan Lee, "The Chinese Fish Culture Classic," trans. Ted S. Y. Moo, www.fao.org.

68 "King Wei of Chi": Ibid.

70 Rather than rotting in jail: Colin E. Nash, *The History of Aquaculture* (Ames, Iowa: Wiley-Blackwell, 2010), p. 13.

70 This marked the birth of underwater polyculture: Ibid.

70 the Indian philosopher Kautilya: Ibid., p. 15.

71 Fishponds covered more than 185,000 acres: Ibid., p. 40.

74 fish farm production topped beef production: Lester Brown, "Farmed Fish Production Overtakes Beef," grist.org, June 12, 2013, https://grist.org.

74 around 70 percent of the salmon: Brian Clark Howard, "Salmon Farming Gets Leaner and Greener," *National Geographic,* March 19, 2014.

74 "Aquaculture has repeated": "About That Salmon," *New York Times,* July 31, 2011.

74 pumping carotenoid and egg yolk additives: "PCBs in Farmed Salmon," ewg.org, July 31, 2013, https://www.ewg.org.

74 Chile, for example: Daniela Guzman, "Great Salmon Escape Threatens to Taint Chile's Fish Farms," *Bloomberg News,* July 9, 2018.

75 The Atlantic Salmon Federation estimated: Food & Water Watch, "The Great Escape: Escapes and Disease Events in Fish Farming," foodandwaterwatch.org, Feb. 2013, https://www.foodandwaterwatch.org.

75 According to Liesbeth van der Meer: Allison Guy, "With Record Antibiotic Use, Concerns Mount That Chile's Salmon Farms Are Brewing Superbugs," oceana.org, Aug. 1, 2016, http://oceana.org.

75   China, for example: "Research Report on China's Trash Fish
     Fisheries Greenpeace East-Asia 2017," Greenpeace (Germany), 2017,
     https://www.greenpeace.de.

76   "I'd rather eat wild cod": Mark Bittman, "A Seafood Snob Ponders
     the Future of Fish," *New York Times,* Nov. 15, 2008.

76   "fundamentally is not sustainable": Michael Pollan, in "Hooked
     on Seafood," episode 3, *Eat: The Story of Food,* National Geographic
     Channel, https://www.nationalgeographic.com.au.

76   In March 2018: John Ryan, "After 3 Decades, Washington State Bans
     Atlantic Salmon Farms," *The Salt,* NPR, March 26, 2018.

76   one out of three fish sold: Kimberly Warner, Patrick Mustain, Beth
     Lowell, Sarah Geren, and Spencer Talmage, "Deceptive Dishes:
     Seafood Swaps Found Worldwide," *Oceana,* Sept. 2016, https://usa
     .oceana.org.

78   "Farmers combine the cultivation": Thierry Chopin, "Integrated
     Multi-Trophic Aquaculture: Ancient, Adaptable Concept Focuses
     on Ecological Integration," *Global Aquaculture Advocate,* March/
     April 2013, https://www2.unb.ca/chopinlab/articles/files/
     Chopin%202013%20GAA%20IMTA.pdf.

78   "As a chef who once quite vociferously preached": Barton Seaver,
     "Opinion: Why You Should Give Farmed Fish a New Look,"
     jamesbeard.org, March 2, 2017, https://www.jamesbeard.org.

79   Compared with their estimate: GGN Certified Aquaculture,
     "Aquaculture and Sustainability II," https://aquaculture.ggn.org.

79   According to a 2017 UN report: United Nations, "World Population
     Prospects: The 2017 Revision," June 21, 2017, https://www.un.org.

80   A recent study in *Nature Communications:* Bernhard Schauberger
     et al., "Consistent Negative Response of US Crops to High
     Temperatures in Observations and Crop Models," *Nature
     Communications* 8 (2017), article no. 13931.

81   To make matters worse: Mark Bittman, *Food Matters: A Guide to
     Conscious Eating with More Than 75 Recipes* (New York: Simon &
     Schuster, 2009).

81   It takes two thousand gallons of water: Ibid.

81   Consumption of fish around the world doubled: Christopher L.
     Delgado, Nikolas Wada, Mark W. Rosegrant, Siet Meijer, and
     Mahfuzuddin Ahmed, "The Future of Fish: Issues and Trends
     to 2020," International Food Policy Research Institute, https://
     ageconsearch.umn.edu/bitstream/15906/1/mi03fu01.pdf.

82   In April 2018: The Canadian Press, "Leading B.C. Chefs Urge
     Province to Terminate Leases for Salmon Farms Opposed by First
     Nations," *Globe and Mail,* April 6, 2018.

82   Twelve months earlier: Courtney Flatt and John Ryan,
     " 'Environmental Nightmare' After Thousands of Atlantic Salmon

Escape Fish Farm," *The Salt,* NPR, Aug. 24, 2017, https://www.npr
.org.

82 One Ark costs twenty million dollars: Neil Ramsden, "Chilean
Firm Seeks Investors to Back Aquaculture-on-a-Vessel Project,"
Undercurrent News, April 3, 2018, https://www.undercurrentnews
.com.

82 "offshore super fish farm": Louis Harkell, "Chinese Consortium
Signs Deal for $1bn Offshore 'Super Fish Farm' Project,"
Undercurrent News, April 18, 2018, https://www.undercurrentnews
.com.

82 Alaska's Senator Lisa Murkowski: Richard Martin, "One Fish, Two
Fish, Strange Fish, New Fish," *bioGraphic,* Feb. 13, 2018, https://www
.biographic.com.

CHAPTER 10: OCEAN RESCUE

95 The Mattabesic tribe: Archibald Hanna, *A Brief History of the Thimble
Islands in Branford, Connecticut* (Hamden, Conn.: Archon Books,
1970), pp. 8–11.

96 Even the best shuckers: Mark Kurlansky, *Cod: A Biography of the Fish
That Changed the World* (New York: Penguin, 1998), p. 182.

96 "Oysters, once plentiful": Quoted in ibid., p. 262.

105 "highly prized and correspondingly expensive": "Oyster Crabs—The
Epicure's Delight," *New York Times,* Nov. 9, 1913.

106 Recent work done by Roger Newell: Roger I. E. Newell, Jeffrey C.
Cornwell, and Michael S. Owens, "Influence of Simulated Bivalve
Biodeposition and Microphytobenthos on Sediment Nitrogen
Dynamics: A Laboratory Study," *Limnology and Oceanography* 47,
no. 5 (2002), pp. 1367–79.

106 A three-acre oyster farm: Richard F. Golen, "Incorporating
Shellfish Bed Restoration into a Nitrogen TMDL Implementation
Plan," Coonamessett Farm, http://www.coonamessettfarm.com/
sitebuildercontent/sitebuilderfiles/Incorporating_Shellfish_Bed_
Restoration_into_Nitrogen_TMDL_Implementation_Plan.pdf.

CHAPTER 11: UP FROM THE DEPTHS

117 Globally, the average body weight: Daniel Pauly and William
W. L. Cheung. "Sound Physiological Knowledge and Principles
in Modeling Shrinking of Fishes Under Climate Change," *Global
Change Biology* 24, no. 1 (2017), pp. e15–e26.

117 Acidification is climbing: IGBP, IOC, SCOR, *Ocean Acidification*

Summary for Policymakers—Third Symposium on the Ocean in a High-CO2 World. International Geosphere-Biosphere Programme, Stockholm, Sweden, 2013, https://news.mongabay.com/2014/02/ocean-acidifying-10-times-faster-than-anytime-in-the-last-55-million-years-putting-polar-ecosystems-at-risk/.

117 The Gulf of Maine: Patrick Whittle, "Gulf of Maine Warming Faster Than 99% of World's Oceans: Study," *CBC,* Sept. 3, 2014, https://www.cbc.ca.

117 And this is all taking place: Amy Novogratz and Mike Velings, "The End of Fish," *Washington Post,* June 3, 2014.

117 In the last ten years alone: Stephen Leahy, "Hidden Costs of Climate Change Running Hundreds of Billions a Year," *National Geographic,* Sept. 7, 2017, https://news.nationalgeographic.com.

118 "quite a shock": Coral Davenport, "Major Climate Report Describes a Strong Risk of Crisis as Early as 2040," *New York Times,* Oct. 7, 2018.

124 Dr. Charles Yarish: Oral history completed by Kendall Barberry and Jill Pegnataro, March 28, 2018, New Haven, Conn.

130 extreme droughts are shriveling farms: Brian Maffly, "Utah Copes with Drying Streams, Dying Animals As Drought Tightens Its Grip—with No Relief in Sight," *Salt Lake Tribune,* Sept. 10, 2018.

130 Recently, two new studies: Georgina Gustin, "Climate Change Could Lead to Major Crop Failures in World's Biggest Corn Regions," *Inside Climate News,* June 11, 2018, https://insideclimatenews.org.

131 With oceans covering 70 percent: "More Than Half of the U.S. Lies Underwater?," *60 Minutes,* May 7, 2015, https://www.cbsnews.com.

131 Professor Ronald Osinga: Roelof Kleis, "Growing Seaweed Can Solve Acidification," Phys.org, Dec. 23, 2010, https://phys.org.

132 Considered the "tree" of coastal ecosystems: Sarah Bedolfe, "Seaweed Could Be Scrubbing Way More Carbon from the Atmosphere Than We Expected," *Oceana Blog,* Oct. 6, 2017, https://oceana.org.

133 "depletes essential oxygen levels": R. A. Duce et al., "Impacts of Atmospheric Anthropogenic Nitrogen on the Open Ocean," *Science* 320 (2008), pp. 893–97.

133 "If done right": Sean Barrett, interview with author, June 28, 2017, New Haven, Conn.

CHAPTER 12: LOOK BACK TO SWIM FORWARD

135 There's even a modern group: "Clam Gardens," *The Clam Garden Network,* https://clamgarden.com/clamgardens/.

135 By the first century B.C.: Mark Kurlansky, *The Big Oyster* (New York: Random House, 2007), pp. 115–17.

135 Elite Romans also kept shellfish: Kenneth F. Kiple and Kriemhild Coneè Ornelas, eds., "Aquatic Animals," *Cambridge World History of Food,* vol. 1 (Cambridge, U.K.: Cambridge University Press, 2000), p. 457.

135 Medieval Europeans continued: Kurlansky, *Big Oyster,* p. 116.

135 By the sixteenth century: Colin E. Nash, *The History of Aquaculture* (Ames, Iowa: Wiley-Blackwell, 2010), p. 44.

136 This area remains the center: Ibid., pp. 56–57.

136 Mussel farming was invented: Maguelonne Toussaint-Samat, "Luxury Foods," in *A History of Food* (Hoboken, N.J.: Wiley, 2008), p. 364.

136 Most scallops eaten in the United States today: Paul Greenberg, "10 Things You Should Know About the American Seafood Supply," *Civil Eats,* July 8, 2014, https: http://civileats.com.

136 In North America, oysters: René E. Lavoie, "Oyster Culture in North America: History, Present and Future," *1st International Oyster Symposium Proceedings,* no. 17, http://www.worldoyster.org /proceeding_pdf/news_17e.pdf.

137 Techniques from the Milford Lab: Nash, *History of Aquaculture,* p. 119.

137 In the little state of Rhode Island: Bren Smith, Sean Barrett, and Paul Greenberg, "What Trump's Budget Means for the Filet-O-Fish," *New York Times,* April 25, 2017.

137 "You've got 1,000 small farms": Quoted in Jeanine Stewart, "Entrepreneur-Driven US Oyster Industry Growing on Farm-to-Table Movement," *Undercurrent News,* Jan. 27, 2018, https://www .undercurrentnews.com.

138 The modern Japanese seaweed industry: Rebecca Rupp, "Like Sushi? Thank a Female Phycologist for Saving Seaweed," *National Geographic,* Feb. 19, 2016, https://www.nationalgeographic.com.

139 Back in Japan: Elisabeth Mann Borgese, *Seafarm: The Story of Aquaculture* (New York: Abrams, 1980).

140 "When the Napoleonic Wars": Claire Eamer, "Seaweed Economics 101: Boom and Bust in the North Atlantic," *Hakai Magazine,* Jan. 5, 2016, https://www.hakaimagazine.com.

140 Massive banks of kelp digesters: Peter Neushul and Lawrence Badash, "Ocean Food and Energy from California Mariculture: An Evaluation of the US Marine Biomass Project, from 1972 to 1986," in *Oceanographic History: The Pacific and Beyond,* ed. Keith R. Benson and Philip F. Rehbock (Seattle: University of Washington Press, 2002).

141 University of Michigan graduate student: Peter Neushul and

Lawrence Badash, "Harvesting the Pacific: The Blue Revolution in China and the Philippines," *Osiris* 13, no. 1 (1998), pp. 186–209.

141 The next wave of U.S. interest: Neushul and Badash, "Ocean Food and Energy from California Mariculture: An Evaluation of the US Marine Biomass Project, from 1972 to 1986."

142 "[A] thousand-acre proof-of-concept kelp farm": George A. Jackson and J. Wheeler North, "Concerning the Selection of Seaweeds Suitable for Mass Cultivation in a Number of Large, Open-Ocean, Solar Energy Facilities ('Marine Farms') in Order to Provide a Source of Organic Matter for Conversion to Food, Synthetic Fuels, and Electric Energy," 1973, U.S. Naval Weapons Center.

142 An illustration of the proposed: *Popular Science* 207, no. 1 (July 1975).

142 scientists like Dr. Michael Neushul: Neushul and Badash, "Ocean Food and Energy," p. 440.

143 "General Electric in December": Jerry Knight, "The $1.2 Million Seaweed Saga, and Other Ideas," *Washington Post,* Feb. 17, 1980.

CHAPTER 14: CENTER OF THE PLATE

162 "What greater joy could be?": Alfred Perceval Graves, *A Celtic Psaltery: Being Mainly Renderings in English Verse from Irish and Welsh Poetry* (New York: F. A. Stokes Company, 1917), p. 21.

166 "harvesting . . . more esteemed ice": Jonathan Miles, "A Brotherhood Formed with Cocktails and Ice," *New York Times,* Dec. 2, 2008.

166 "has no interest in the veggie burger": Tejal Rao, "A Pastry Chef Masters the Veggie Burger," *New York Times Style Magazine,* July 15, 2014.

171 "shockingly delicious": Sam Sifton, interview with author, Sept. 16, 2016.

176 "the food pyramid": Dana Goodyear, *Anything That Moves: Renegade Chefs, Fearless Eaters, and the Making of a New American Food Culture* (New York: Riverhead, 2013), p. 49.

CHAPTER 15: SEA GREENS FOR THE MASSES

178 Between 2011 and 2014: Lindsay Quinn, "The Great Kalespiracy," *The Hustle,* May 9, 2018, https://thehustle.co.

179 His nonprofit helped serve kale: Rebekah Kebede, "Who Owns Kale?," *National Geographic,* Aug. 10, 2016, https://www.nationalgeographic.com.

179 kale was surfacing as a hot baby name: Bruce Horovitz, "Vegetables Shift to Center of the Plate," *USA Today*, Nov. 9, 2013.

183 "one of the many shelf-stable cans": Tejal Rao, "A Pastry Chef Masters the Veggie Burger," *New York Times Style Magazine*, July 15, 2014.

183 a glowing review: *Soybean Digest*, Dec. 1942, p. 8.

185 When the amount of residue left: Laura Sesana, "EPA Raises Levels of Glyphosate Residue Allowed in Food," *Washington Times*, July 5, 2013; "Pesticide Tolerances: Glyphosate," July 1, 2013, Environmental Protection Agency, https://www.regulations.gov.

185 This increased limit remains: Kathryn Z. Guyton et al., "Carcinogenicity of Tetrachlorvinphos, Parathion, Malathion, Diazinon, and Glyphosate," *Lancet Oncology* 16, no. 5 (2015), pp. 490–91.

185 Recently leaked FDA emails: Carey Gillam, "Weedkiller Found in Granola and Crackers, Internal FDA Emails Show," *The Guardian*, April 30, 2018.

185 "Sequoia of the Sea": Dorte Krause-Jensen and Carlos M. Duarte, "Substantial Role of Macroalgae in Marine Carbon Sequestration," *Nature Geoscience* 9 (2016), pp. 737–42.

186 In the eighth century: Hassan Khalilieh, "A Glimpse on the Uses of Seaweeds in Islamic Science and Daily Life During the Classical Period," *Arabic Sciences and Philosophy* 16, no. 1 (March 2006), pp. 91–101.

186 Even the best land-based farms: Tatiana Schlossberg, "Fertilizers, a Boon to Agriculture, Pose Growing Threat to U.S. Waterways," *New York Times*, July 27, 2017.

187 archaeozoologist Ingrid Mainland: Slow Food in the UK, The UK Ark of Taste & Chef Alliance Programme, Ark of Taste Producers, "North Ronaldsay Sheep," http://www.slowfood.org.uk.

188 A report commissioned by the European Union: Pete Harrison, "Once-Hidden EU Report Reveals Damage from Biodiesel," *Reuters*, April 21, 2010.

189 Scientists at the University of Indiana: Igor Kovalenko et al., "A Major Constituent of Brown Algae for Use in High-Capacity Li-Ion Batteries," *Science* 334 (2011), pp. 75–79.

189 The DOE estimates: Renee Schoof, "Seaweed in the Tank? Company Turns to Aquaculture for Ethanol," McClatchy Newspapers, Jan. 19, 2012.

189 The world's energy needs could be met: Tim Flannery, *Sunlight and Seaweed: An Argument for How to Feed, Power and Clean Up the World* (Melbourne: Text Publishing, 2017), p. 141.

189 "equivalent of striking oil": University of California, Berkeley,

"Common Algae Can Be Valuable Source of Hydrogen Fuel," ScienceDaily, Feb. 23, 2000.

190 Locally, we've been pushing for an expansion: Christine Buckley, "Seaweed: The New Trend in Water Purification," *UConn Today,* July 27, 2011, http://today.uconn.edu/.

CHAPTER 17: SWIMMING WITH SHARKS

218 "If I could buy kelp futures": Dina Spector, "Forget Tuna: These Are the Seafoods We'll Be Eating in the Future," *Business Insider,* July 8, 2014.

221 "We deforest the land": Fred Imbert, "Capitalism Is Killing the Planet and Needs to Change, Says Investor Jeremy Grantham," CNBC, June 13, 2018, https://www.cnbc.com.

CHAPTER 18: NEWFOUNDLAND, TAKE ME HOME

234 "You were walking around": "From This Place: Our Lives on Land and Sea," Ongoing Exhibition, The Rooms Museum, St. John's, Newfoundland, Canada.

236 He used a homespun Newfoundland pickling recipe: Sam McNeish, "Torbay Farmer Harvesting New Bounty—Seaweed," *The Telegram,* March 12, 2018, http://www.thetelegram.com.

CHAPTER 19: A SHARED VISION

246 "People need a new jacket": Yvon Chouinard, on twitter.com /PatagoniaProv, March 11, 2018, https://twitter.com/PatagoniaProv /status/972962569244692480.

246 According to Birgit Cameron: Katie O'Reilly, "Patagonia Intends to Help Solve Climate Change with Snack," *Sierra,* March 29, 2017, https://www.sierraclub.org.

248 New Communities: http://www.newcommunitiesinc.com.

249 Food Warriors Youth Development Program: Hannah Wallace, "Malik Yakini of Detroit's Black Community Food Security Network," *Civil Eats,* Dec. 19, 2011, https://civileats.com.

250 "Most ownership structures": Adele Peters, "This Company Pioneered a New Business Structure to Preserve Its Mission," *Fast Company,* July 13, 2018, https://www.fastcompany.com.

250 "There is an abundance": Capital Institute, "Think Tank Led by

Former JPMorgan Managing Director Launches Regenerative Capitalism Framework at Yale University," *PR Newswire,* April 22, 2015, https://www.prnewswire.com.

251 As Dan Barber, a hero of the organic food movement: Dan Barber, *The Third Plate* (New York: Penguin, 2014), p. 175.

253 It was too low: "ABO Scores Historic Victory for Carbon Utilization," Algae Biomass Organization, Feb. 9, 2018, https://algaebiomass.org.

255 Inventories and sales: Demian A. Willette et al., "Using DNA Barcoding to Track Seafood Mislabeling in Los Angeles Restaurants," *Conservation Biology* 31, no. 5 (2017), pp. 1076–85.

CHAPTER 20: FADE TO BLUE

258 Food is the largest private employment sector: Amy Frykholm, "Food Justice Is About Workers," *Center for Good Food Purchasing,* Jan. 17, 2017, https://goodfoodpurchasing.org.

Index

A NOTE ABOUT THE AUTHOR

Bren Smith is a former commercial fisherman turned ocean farmer who pioneered the development of restorative 3D ocean farming. Born and raised in Newfoundland, he left high school at the age of fourteen to work on fishing boats from the Grand Banks to the Bering Sea. He has been the owner of Thimble Island Ocean Farm for fifteen years and started the nonprofit GreenWave to train a new generation of ocean farmers. He was named one of *Rolling Stone*'s 25 People Shaping the Future and his 3D ocean farm was selected as one of *Time* magazine's Best Inventions of 2017. Bren's work has been profiled by *The New Yorker*, *National Geographic*, CNN, *60 Minutes*, *Scientific American*, *Bon Appétit*, and *Outside*, among others. Bren lives in Fair Haven, Connecticut, in an 1875 oyster captain's house, with his wife, his cat Fisher, and his Newfoundland Juniper.

A NOTE ON THE TYPE

This book was set in Legacy Serif. Ronald Arnholm (b. 1939)
designed the Legacy family after being inspired by the 1470
edition of *Eusebius* set in the roman type of Nicolas Jenson.
This revival type maintains much of the character of the
original. Its serifs, stroke weights, and varying curves give
Legacy Serif its distinct appearance. It was released by the
International Typeface Corporation in 1992.

Composed by North Market Street Graphics,
Lancaster, Pennsylvania

Printed and bound by Berryville Graphics,
Berryville, Virginia